# PIPE FITTINGS



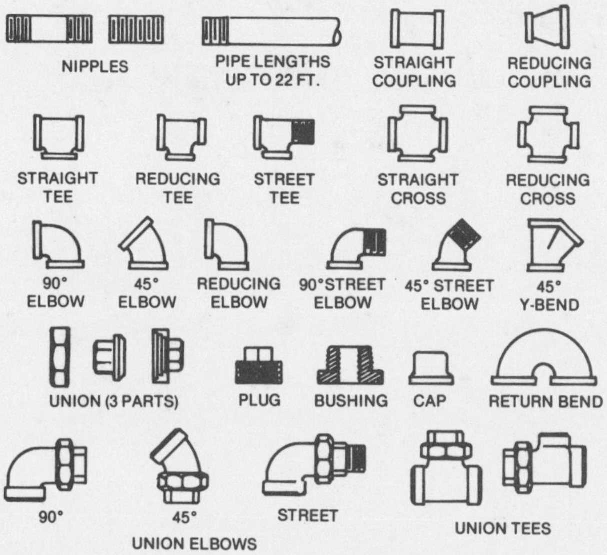


Here are the common steel pipe fittings. Nipples are simply short lengths of pipe threaded on both ends. Reducing fittings join two different sizes of pipe.

Compression fittings of the flared-tube type are the easiest for the novice to handle when working with copper tubing.

| STANDARD STEEL PIPE (All Dimensions in Inches) | | | | | |
|---|---|---|---|---|---|
| Nominal Size | Outside Diameter | Inside Diameter | Nominal Size | Outside Diameter | Inside Diameter |
| 1/8 | 0.405 | 0.269 | 1 | 1.315 | 1.049 |
| 1/4 | 0.540 | 0.364 | 1 1/4 | 1.660 | 1.380 |
| 3/8 | 0.675 | 0.493 | 1 1/2 | 1.900 | 1.610 |
| 1/2 | 0.840 | 0.622 | 2 | 2.375 | 2.067 |
| 3/4 | 1.050 | 0.824 | 2 1/2 | 2.875 | 2.469 |

## SQUARE MEASURE

144 sq in = 1 sq ft
9 sq ft = 1 sq yd
272.25 sq ft = 1 sq rod
160 sq rods = 1 acre

## VOLUME MEASURE

1728 cu in = 1 cu ft
27 cu ft = 1 cu yd

## MEASURES OF CAPACITY

1 cup = 8 fl oz
2 cups = 1 pint
2 pints = 1 quart
4 quarts = 1 gallon
2 gallons = 1 peck
4 pecks = 1 bushel

## WOOD SCREWS

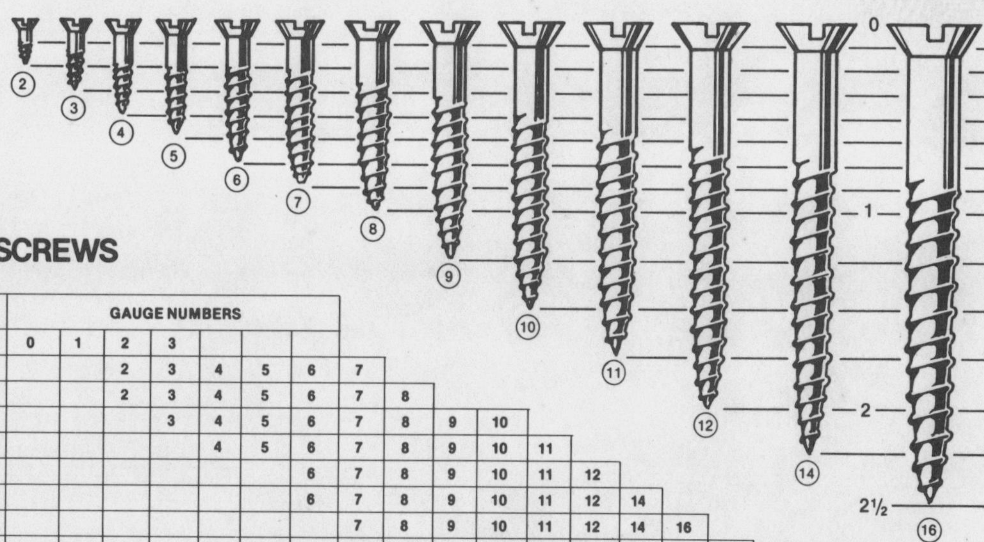


| LENGTH | GAUGE NUMBERS | | | | | | | | | | | | | | | | |
|---|---|---|---|---|---|---|---|---|---|---|---|---|---|---|---|---|---|
| 1/4 INCH | 0 | 1 | 2 | 3 | | | | | | | | | | | | | |
| 3/8 INCH | | | 2 | 3 | 4 | 5 | 6 | 7 | | | | | | | | | |
| 1/2 INCH | | | 2 | 3 | 4 | 5 | 6 | 7 | 8 | | | | | | | | |
| 5/8 INCH | | | | 3 | 4 | 5 | 6 | 7 | 8 | 9 | 10 | | | | | | |
| 3/4 INCH | | | | | 4 | 5 | 6 | 7 | 8 | 9 | 10 | 11 | | | | | |
| 7/8 INCH | | | | | | | 6 | 7 | 8 | 9 | 10 | 11 | 12 | | | | |
| 1 INCH | | | | | | | 6 | 7 | 8 | 9 | 10 | 11 | 12 | 14 | | | |
| 1 1/4 INCH | | | | | | | | 7 | 8 | 9 | 10 | 11 | 12 | 14 | 16 | | |
| 1 1/2 INCH | | | | | | | 6 | 7 | 8 | 9 | 10 | 11 | 12 | 14 | 16 | 18 | |
| 1 3/4 INCH | | | | | | | | | 8 | 9 | 10 | 11 | 12 | 14 | 16 | 18 | 20 |
| 2 INCH | | | | | | | | | 8 | 9 | 10 | 11 | 12 | 14 | 16 | 18 | 20 |
| 2 1/4 INCH | | | | | | | | | | 9 | 10 | 11 | 12 | 14 | 16 | 18 | 20 |
| 2 1/2 INCH | | | | | | | | | | | | | 12 | 14 | 16 | 18 | 20 |
| 2 3/4 INCH | | | | | | | | | | | | | | 14 | 16 | 18 | 20 |
| 3 INCH | | | | | | | | | | | | | | | 16 | 18 | 20 |
| 3 1/2 INCH | | | | | | | | | | | | | | | | 18 | 20 | 24 |
| 4 INCH | | | | | | | | | | | | | | | | 18 | 20 | 24 |

WHEN YOU BUY SCREWS, SPECIFY (1) LENGTH, (2) GAUGE NUMBER, (3) TYPE OF HEAD—FLAT, ROUND, OR OVAL, (4) MATERIAL—STEEL, BRASS, BRONZE, ETC., (5) FINISH—BRIGHT, STEEL BLUED, CADMIUM, NICKEL, OR CHROMIUM PLATED.

# In this volume . . .

**MODELMAKERS ATTENTION!** Here's a magnificent project in gleaming brass. It's a finely detailed model of a gun from the main battery of the U.S. frigate *Constitution*. You'll be proud to display it on your bookshelf or desk. The plans are on page 574.

**SHOWN BELOW AND AT RIGHT,** according to one expert, is the basic equipment for a canoe-camping expedition. You'll find it listed and described in the article starting on page 581. Of course, one man's wilderness may be another man's back yard, so you may disagree with some of the items. Even so, we'll bet you find the article—and the list—fascinating!

**NO TWO CANDLES** molded in snow are ever alike, just as no two snow flakes are ever alike. Learn how to make these beautifully delicate free-form candles in the article on page 560.

**THROW A TAILGATE PARTY!** Above and at left you see our portable compartmented organizer, nicely insulated and beautiful to behold—a chuck wagon with a college degree. Side sections can be separated from the insulated center section for individual use. A hinged door on the left section opens to reveal a complete portable bar with mixing equipment and storage for all popular bottle sizes. The center section's door drops down to become a service counter. Tableware and utensils are stored in the right section. Of course you don't need a football game to enjoy a picnic from the chuck wagon. See page 494.

**HERE'S ONE MAN'S VERSION** of the perfect home workshop. Not everyone, of course, can afford such an elaborate shop. You'll find fine ideas for laying out a shop of any size on page 3142 of Vol. 20.

# Look what you'll find in other volumes!

**EVERY MOTHER'S FEAR** is to awaken and smell smoke. What she does immediately after evacuating the kids and calling the fire department may save her home. You'll find a complete rundown on extinguishers and how to use them in Vol. 8, page 1218.

**A MULTILEVEL DECK** makes good sense because it reduces the amount of furniture you'll need when you entertain a crowd. You'll find suggestions for building this, or any other deck, in a fine article on page 920 of Vol. 6.

**THIS ONCE WAS A GARAGE!** The charming cottage shown at left started out as a detached building. Imaginative use of space and materials transformed it into the comfortable guest house shown above. See Vol. 9, page 1316.

**CHECK THESE TOYS** for child appeal! There's a swallowing whale, a woodblock train, a puffing engine, a giraffe that holds clothes, and a circus-wagon toy box. You'll find plans and patterns for all of them in Vol. 19, page 2914.

**FOR NOSTALGIA BUFFS,** here's a classic rolltop desk you can build yourself. Talk about turn-of-the-century charm! This fine piece of furniture would cost $1300 or more if you bought it as an antique. See the step-by-step construction plans on page 954 of Vol. 6

**CHOOSE JUST** the right shrubs and ground covers so you'll have year-round visual interest but no demanding upkeep. In Vol. 11 on page 1679 you'll find charts, diagrams and much detailed information on the amazing variety of flowering and evergreen shrubs available to you.

**THE CANTILEVERED** steps seem made for that hillside. In "Handsome garden paths and walkways," Vol. 14, page 2084, we include ideas for using inlaid stones, concrete slabs, paths with built-in greenery, an asphalt path, and several decorative effects for concrete.

Popular Mechanics

# do-it-yourself encyclopedia

in 20 volumes

a complete how-to guide for the homeowner, the hobbyist—
and anyone who enjoys working with mind and hands!

All about:

home maintenance
home-improvement projects
wall paneling
burglary and fire protection
furniture projects
finishing and refinishing furniture
outdoor living
home remodeling
solutions to home problems
challenging woodworking projects
hobbies and handicrafts
model making
weekend projects
workshop shortcuts and techniques
hand-tool skills
power-tool know-how
shop-made tools
car repairs
car maintenance
appliance repair
boating
hunting
fishing
camping
photography projects
radio, TV and electronics know-how
clever hints and tips
projects just for fun

volume 4

ISBN 0-87851-069-9

Library of Congress Catalog Number 77 84920



The Hearst Corporation cannot assume responsibility for proper application of techniques, implementations or proper and safe functioning of any of the manufactured or reader-built items, products or projects resulting in whole or in part from all or any part of any plans, prints, projects or instructional materials published in this volume.

The Hearst Corporation cannot assume responsibility for the continuing availability of any special parts listed in articles in this volume.

MANUFACTURED IN THE UNITED STATES OF AMERICA

# contents

VOLUME 4

# 'Instant' furniture from kitchen cabinets

By JIM LISTON

STUDY CENTER

WET BAR

ROOM DIVIDER

■ IN ALMOST NO TIME you can build this entertainment wall in living room, family room or den—even if you're not a great carpenter. Yours will look every bit as good because you use ready-built modules—today's furniture-quality kitchen cabinetry.

Even experienced handymen like to work with ready-mades; with everything true, glued, hinged and finished, they're pretty hard to beat at today's prices for hardwood.

The free-standing entertainment wall, and other pieces suggested above left and on the following pages, can be built with any kitchen storage units. We chose Olde Dominion, an all-oak line by Boise Cascade Kitchen Cabinet Div. These cabinets have a hand-rubbed wood-grain finish, and the construction equals that of top-quality furniture.

Like playing with building blocks, you simply stack the cabinets; their weight alone is suffi-

## Entertainment wall

**Basic units**
5 15 x 30-in. single wall cabinets
3 30 x 30-in. double wall cabinets
2 15 x 30-in. double wall cabinets

**Large open-back boxes (½" plywood)**
6 pcs. ½ x 12 x 30-in.
6 pcs. ½ x 12 x 29⅛-in.
2 pcs. ½ x 11½ x 29-in.

**Small open-back boxes (½" plywood)**
4 pcs. ½ x 12 x 30⅛-in.
4 pcs. ½ x 12 x 14-in.
4 pcs. ½ x 11½ x 14-in.

**Cost** (cabinets only): Olde Dominion, about $600; Woodcrest, approximately $400.

SEE ALSO
**Bars . . . Bookcases . . . Desks . . . Drawers . . .**
**Hi-fi centers . . . Modular furniture . . .**
**Room dividers . . . Sewing centers . . .**
**Storage walls . . . Study centers . . .**
**Theaters, home . . . Varnish**

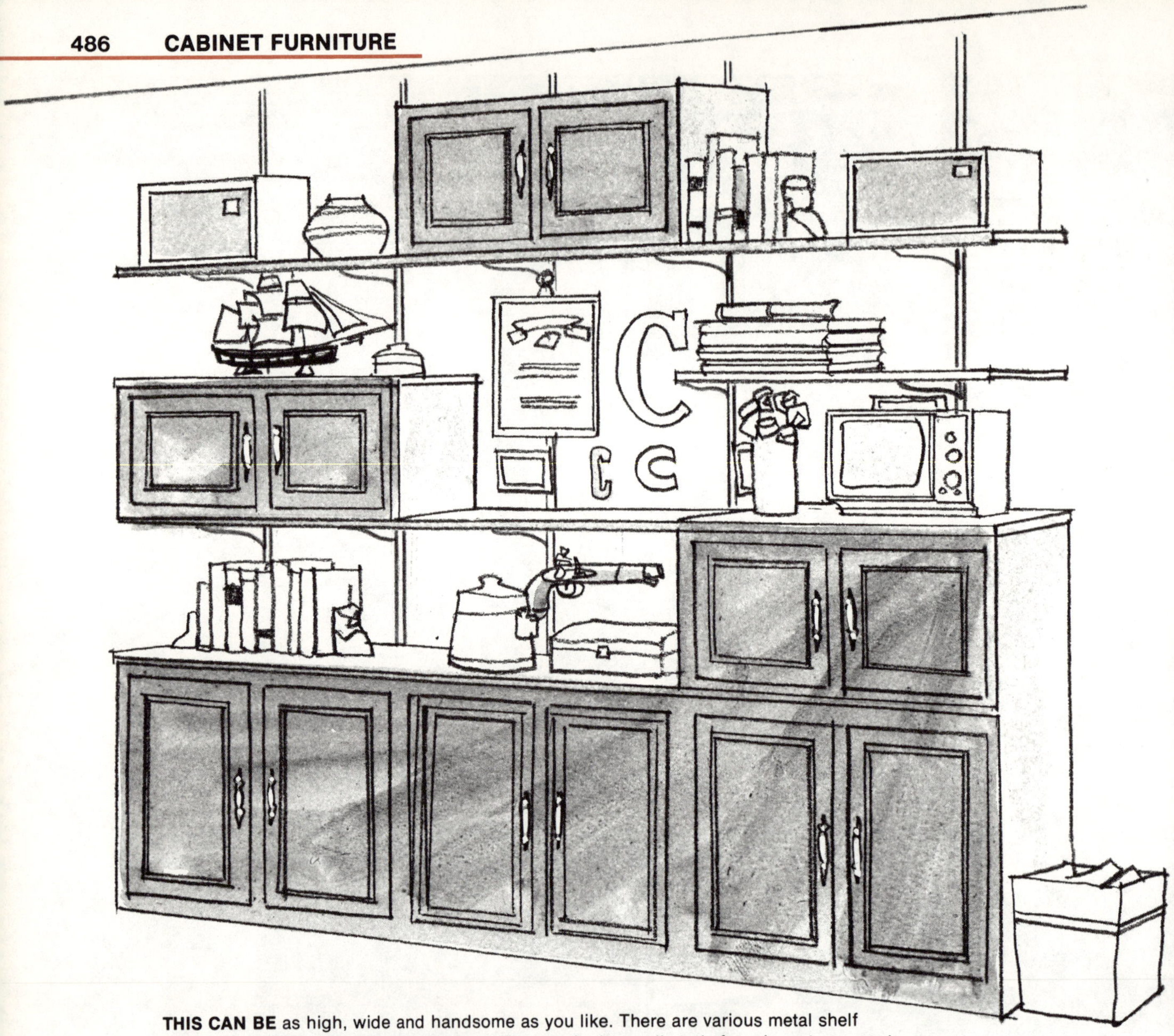

**THIS CAN BE** as high, wide and handsome as you like. There are various metal shelf systems available. Shelf standards should be attached to wall studs for adequate support. Shelving can be finished to match cabinets or painted a bright contrasting color.

**Storage Wall**

3-30x30-in. wall cabinets
3-15x30-in. wall cabinets
3-1x12x90-in. shelving
2-1x12x30-in. shelving
5-6-ft. shelf standards
10-12-ft. shelf brackets

cient to keep them standing securely but it might be wise to anchor the upper cabinets. In the case of the entertainment wall, 10 standard-size, kitchen cabinets are grouped around five easy-to-make, open-back boxes of modular size, which you glue and nail together from ½-in. plywood, then paint.

The big plus in using kitchen cabinets is economy and availability. Lumber yards and kitchen remodeling dealers can offer almost immediate delivery, but there's often a long wait for furniture.

Decorate with Cabinets, *an idea booklet on how to build furniture from cabinets, is available from Boise Cascade Kitchen Cabinet Div., Box 514, Berryville, Va. 22611. Enclose 50 cents to cover postage and handling.*

**TABLES** from wall cabinets suggest all kinds of possibilities. Top them with self-edged, plastic-laminated plywood. The plastic can be wood-grained, white or a bright accent color. The approximate costs for the cabinets only: Olde Dominion, about $165; Estate Oak, about $100.

**FOR STUDENT,** menu-planner or budget-keeper, this good-looking desk can be made quickly and economically from kitchen-cabinet bases and a plastic-covered plywood top. Bases are available with deep drawers that take standard file folders. Cost (cabinets only): Olde Dominion, about $225; and Woodcrest, about $150.

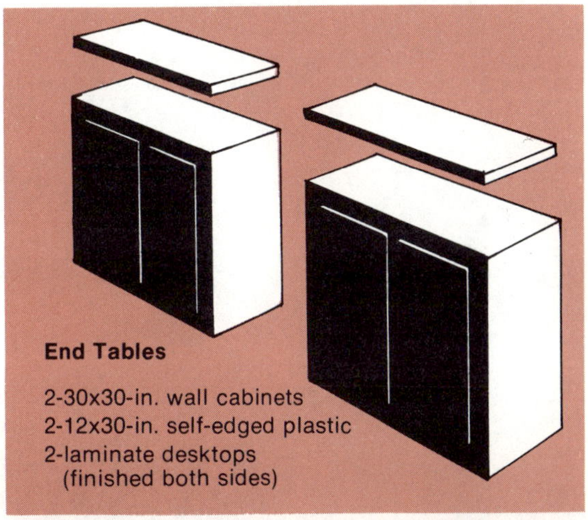


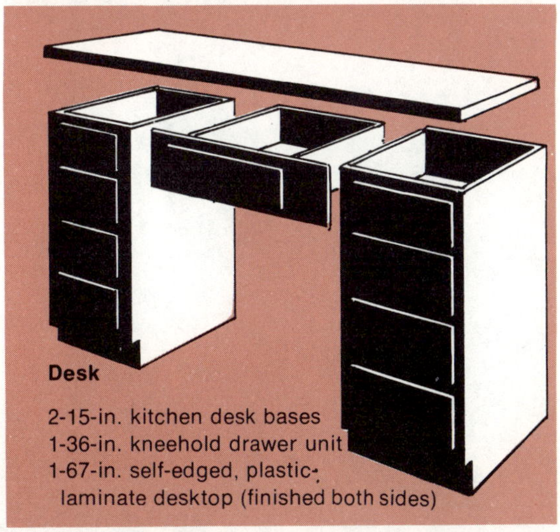

# Build a great wall of sound

■ WHAT TURNS YOU on to a stereo system? For most people it's a combination of good sound and good looks. The cost is generally the only hangup. The way to beat the high price tags is to build the cabinets yourself and sink your money into individual components.

We made the cabinets from birch-veneer plywood and kept the design clean and simple so it would look good without overpowering the room. Speaker construction is not complicated, but make sure you follow the dimensions carefully. A computer picked them as the best possible proportions from thousands of combinations.

Although a mitered corner joint might have

looked best (since we cut the panels from the same sheet and matched the grain at the joints) we chose a half-lap with a solid birch band to cover the plywood end grain. This joint is easy to make, protects the veneered edges and can stand up to the powerful vibrations generated by the 12-in. woofer.

On the inside, glue and screw ¾-in. plywood strips to support all the interior joints. Cut the front and back panels after assembling the frame.

The sleek cabinet is designed to show off your components and hide all your storage. Starting from the bottom, there are two, full-length shelves for record storage. The ¼-in.-thick dividers serve as a trusslike support system for the weight above. Next comes a 6-in.-high, full-depth storage area. The fold-down door can serve as a work surface for sorting or splicing tapes. Inside there's room for headphones, microphones, cleaning equipment, instruction booklets and other small items.

Above this is the main component board. Glue and clamp this carefully. You can add glue blocks and screws from the inside for extra support.

To increase visibility, we surfaced the interior

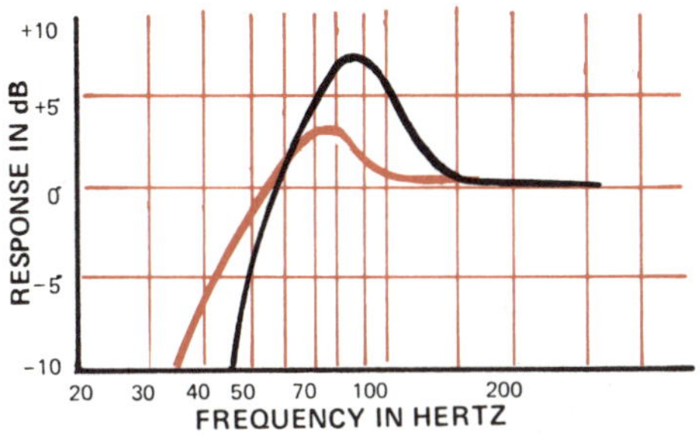


**THE COMPUTER'S** best effort produced the "A" curve. Even a small error in the dimensions of the port would change the smooth, deep bass of the curve into the falsely peaked bass of curve "B," a design the computer rejected. We used modern Electro-Voice components in the speakers.

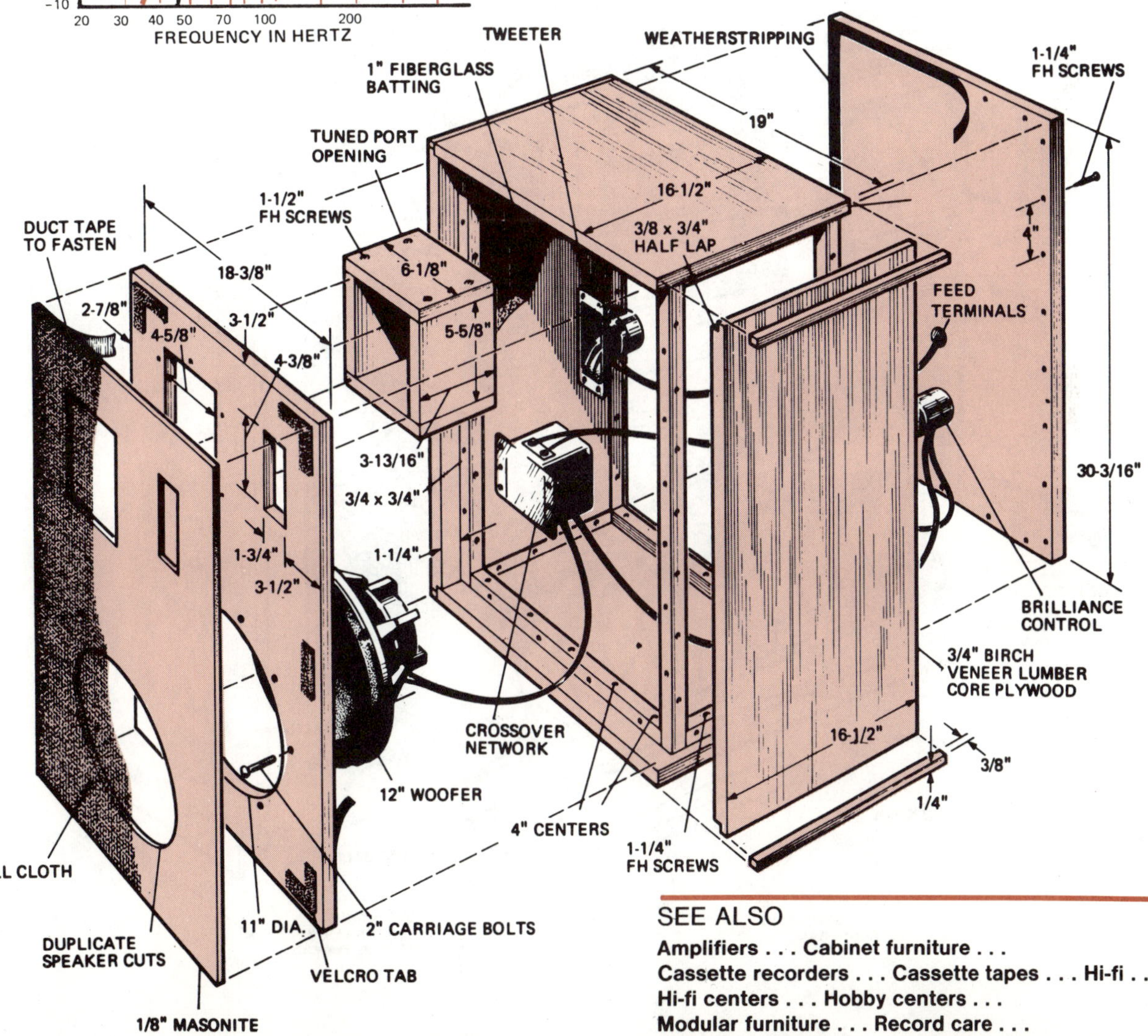


SEE ALSO

**Amplifiers . . . Cabinet furniture . . . Cassette recorders . . . Cassette tapes . . . Hi-fi . . . Hi-fi centers . . . Hobby centers . . . Modular furniture . . . Record care . . . Television antennas . . . Theaters, home**

of the turntable opening with mirrors and added a pencil fluorescent fixture. This elegant extra makes the whole unit look like a million dollars at night. Above the components is another 6-in. storage section. But behind this door we mounted a series of cassette holders made by 3M. On each one, a touch of the button releases a spring-loaded drawer and a tape pops out, ready to load. Above the tape door (which you can hinge to fold up or down) is another shelf for record storage.

You can order the Electro-Voice components shown from *Popular Mechanics* for about $314 for the stereo set. Query first regarding current price. Address your query to *Popular Mechanics,* Home & Shop Dept., Box 1014, Radio City Station, New York, NY 10019.

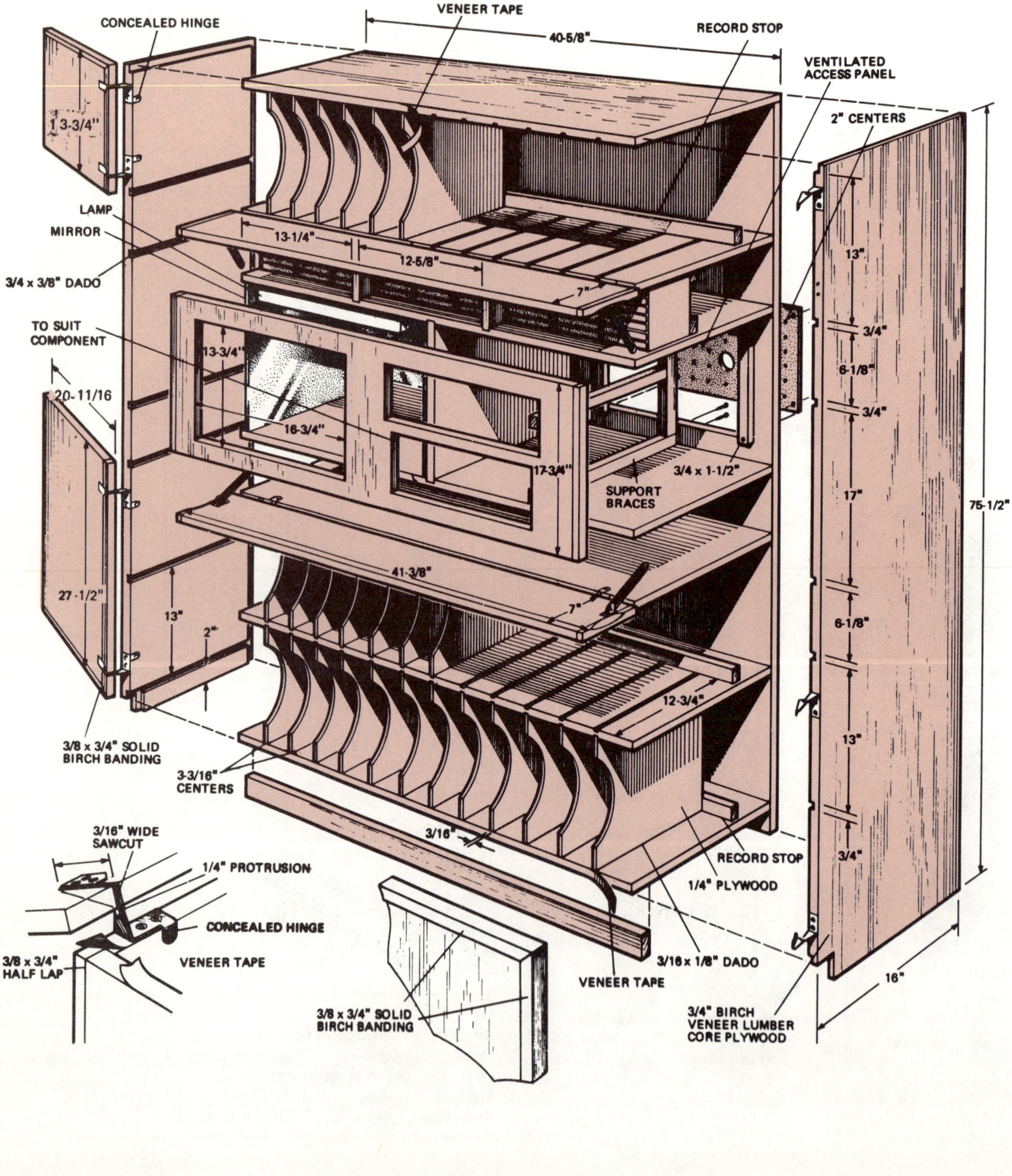

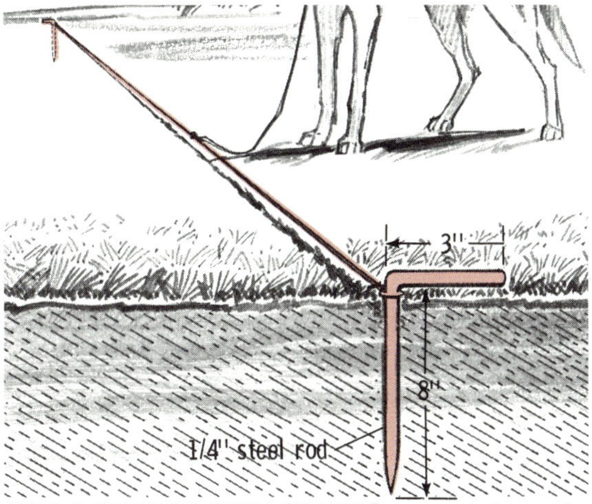


**TWO L-SHAPED IRON STAKES** passed through eyes in the ends of a galvanized wire and then driven into the ground, make a non-trip-over tether for your dog. The dog's leash is snapped to the wire.—*John Krill.*

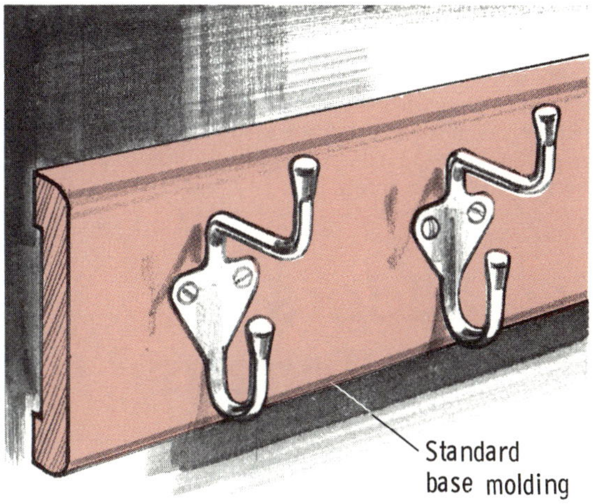


**WHEN ADDING COAT HOOKS** to a closet wall, a strip of door casing makes a neat rail on which to screw them. Besides being nicely shaped, the casing provides a solid surface for screws.—*Daniel Bousha.*

**WHEN WATER MUST BE CARRIED** in an open container in the car, it will be sloshproof if you first line the pail with a plastic bag. After filling, bag is tied at the top with a bit of string.—*Andrew Vena.*

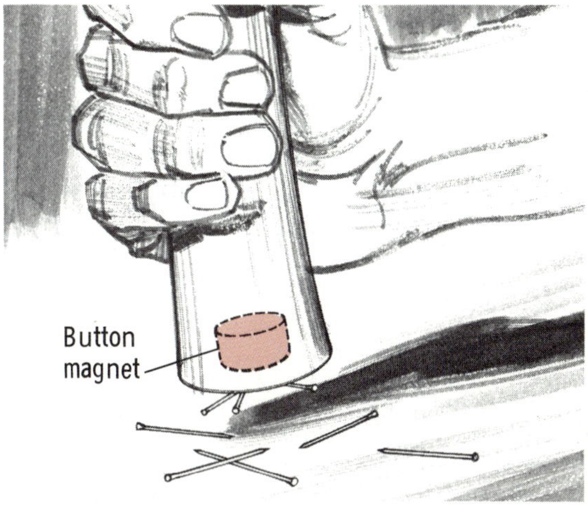


**PICKING UP TINY NAILS** is no problem if you embed a small button magnet in your hammer handle. Drill a hole slightly smaller than the magnet and tap the latter in place.—*Joseph Braunstein.*

**YOU CAN SAVE YOURSELF STEPS** when you make use of a grass catcher to increase the capacity of your wheelbarrow. Use pieces of wire to hang it between the handles of the barrow.—*Victor H. Lamoy.*

### 1. Unused soffit space

**THE DEAD SPACE** going to waste within the soffits over kitchen cabinets offers handy storage for items used only on occasion like Christmas dishes and glassware. Make the opening in the plasterboard as large as space permits, then install short studs at the sides to take a lip-type door. Fasten door with a magnetic catch.

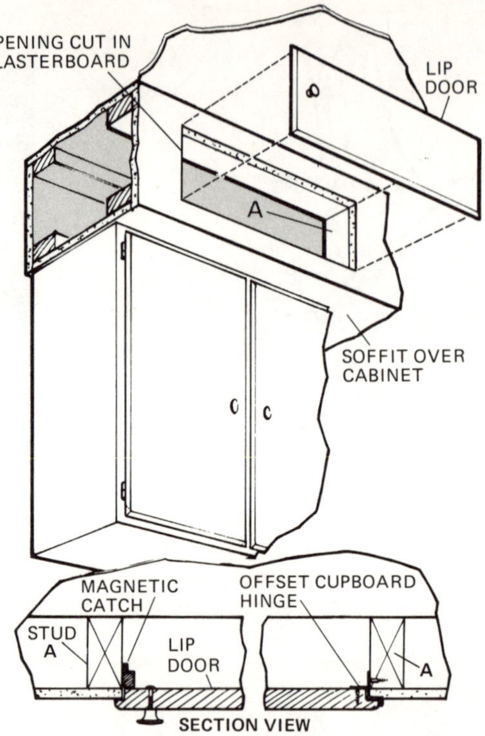


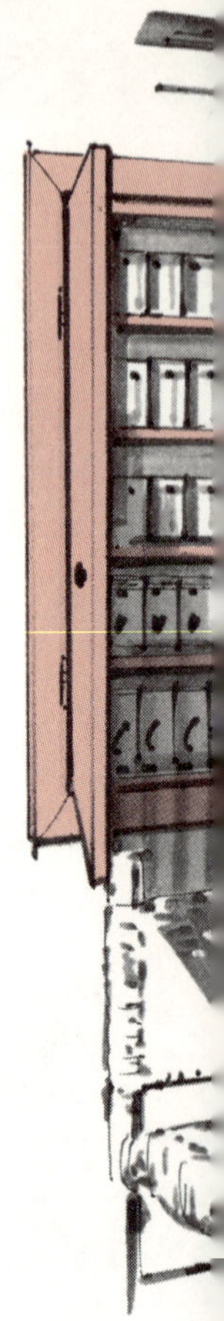

# Space adders for crowded kitchens

**By WAYNE C. LECKEY**

■ IF YOU'RE HURTING for extra cupboard space and see no way of gaining more, chances are you have overlooked potential storage space that's actually there and going to waste. In some cases it's dead space within a kitchen wall and the soffits over cabinets; in others it's space under wall cabinets and on the surface of a door.

The four space adders shown here show how you can put this waste space to work by building in or adding on shallow cabinets to give extra storage for canned goods, spices and items you don't often use.

SEE ALSO

**Kitchen accessories . . . Kitchens . . .**
**Oven hoods . . . Storage walls**

### 2. Piggyback door storage

**HANG A SHALLOW** cabinet fitted with shelves and a door on the inner surface of a hollow flush door and you'll gain more storage space for can and package goods. Make the cabinet of 1 x 4s, add a dowel railing across each shelf to hold cans in place and hang it with L-brackets and Molly bolts.

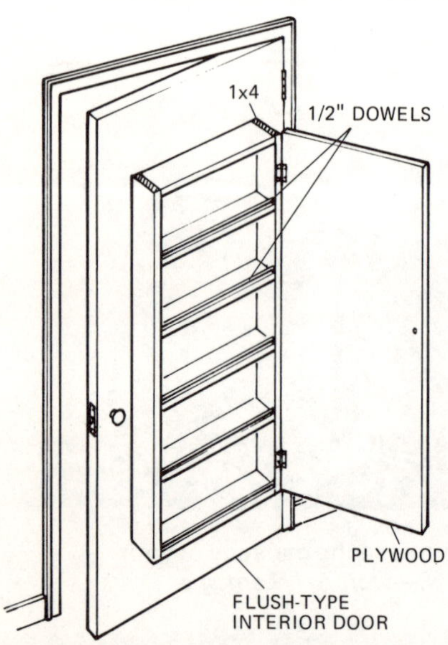


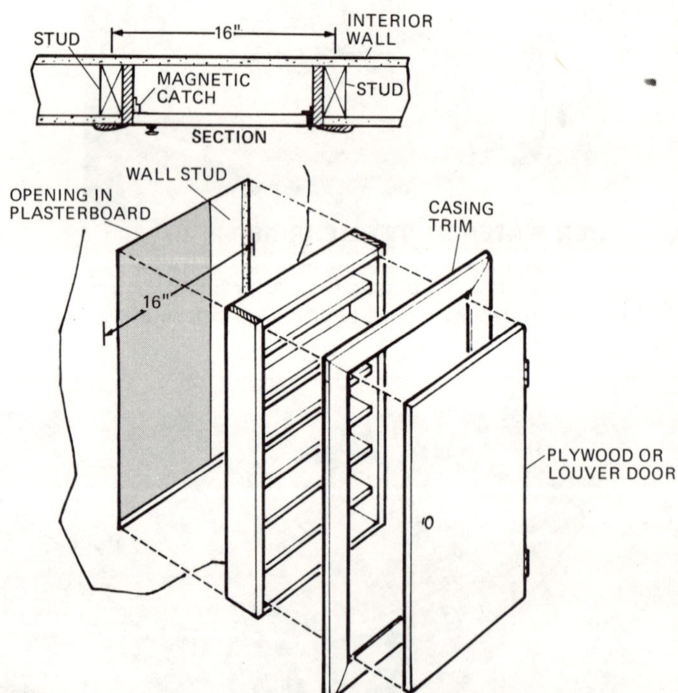


### 3. Wall niche for canned goods

**HOLLOW SPACE** in an interior wall offers storage galore for canned goods. The 3½-in. width of the studs is ample to accept row upon row, one-can deep. The cabinet is made to fit between a pair of studs. The door can be fitted flush with the wall for soup-size (2¼-in.) cans, or it can be brought out an additional ¾ in. by a deeper frame to accept 3-in.-dia. cans. See that the wall is free of pipes and wires.

## 4. Under-cabinet storage

**WITHOUT SACRIFICING** counter space, extra storage space can be gained by hanging a row of 4-in. binlike drawers from the underside of your wall cabinets. They're handy for keeping spices, napkins and the like at your fingertips. Each hinged bin pulls out like a drawer and tips down at an angle for full access to contents. The bins ride in a frame assembly attached to the cabinets with screws in counterbored holes. Stop blocks in the grooves prevent the bins from being pulled all the way out.

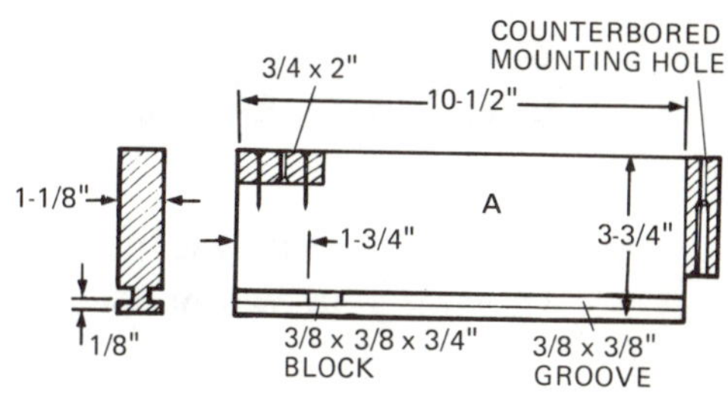


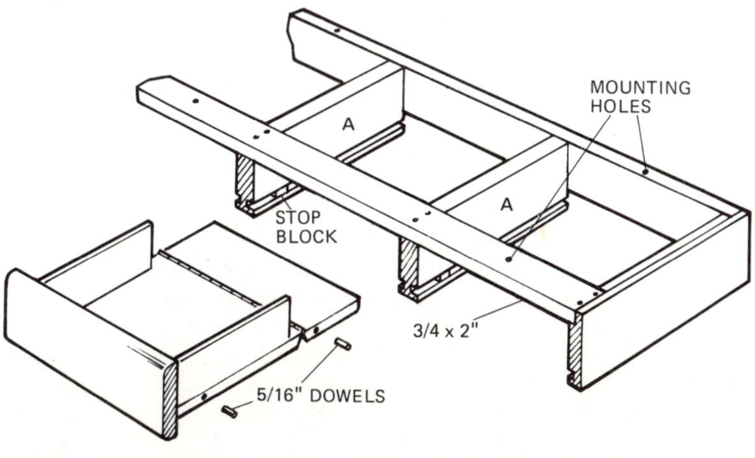


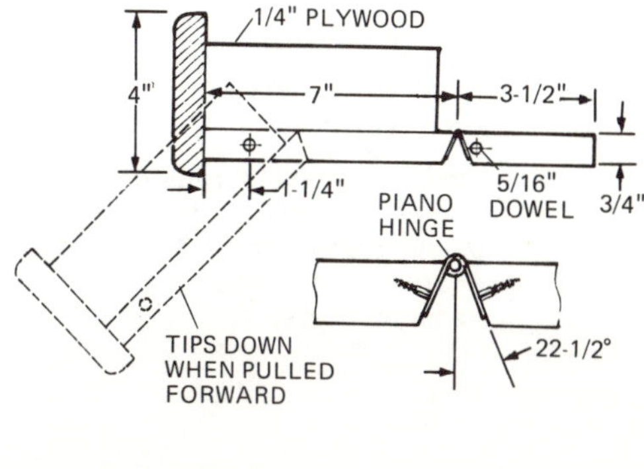

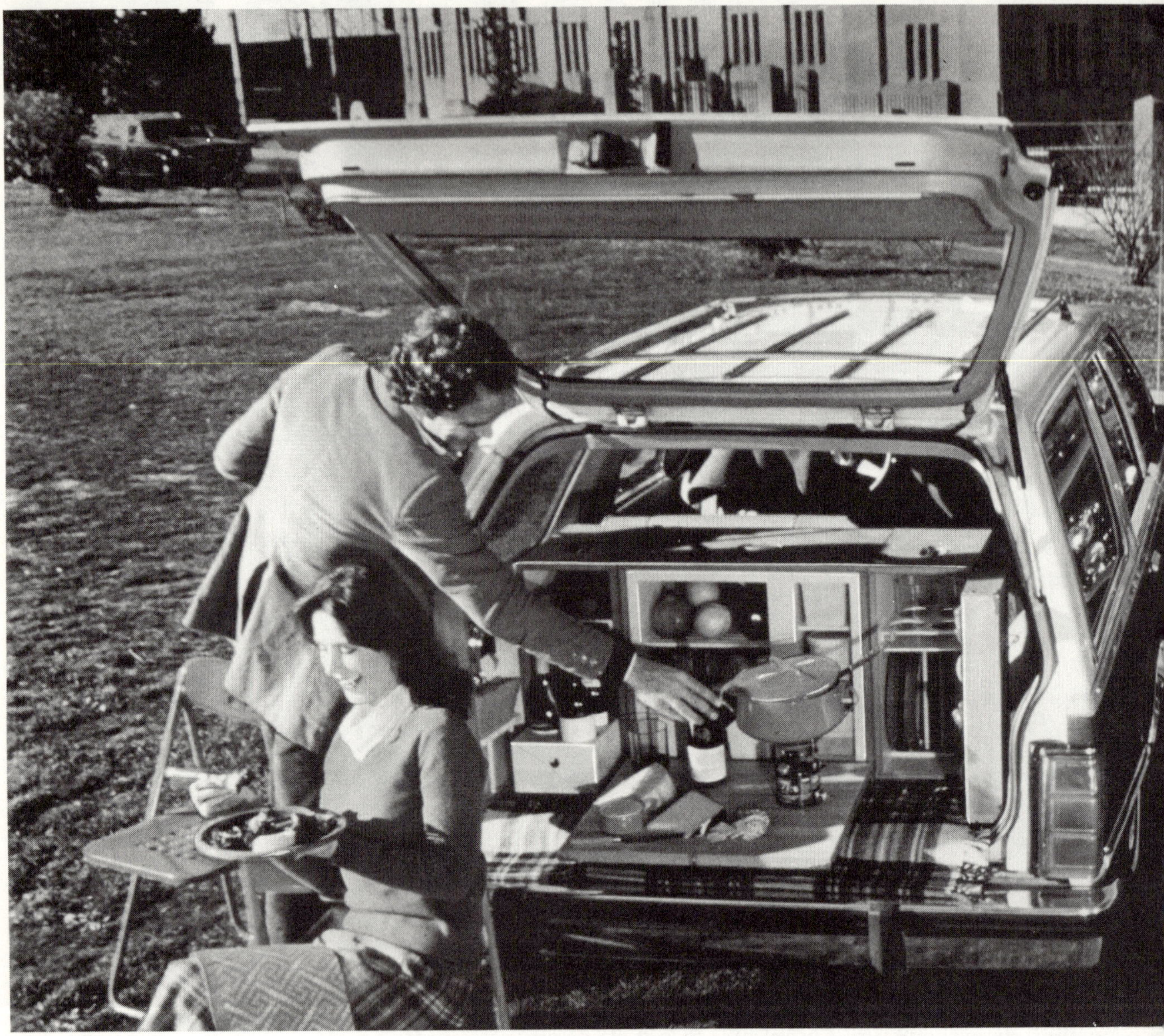

# Come to a tailgate party!

**Sandwiches and beer just aren't a class act at the stadium any more. To cater your own parking-lot buffet, build this insulated compartmentalized organizer**

SEE ALSO

**Bars . . . Cabinet furniture . . . Camping . . . Chests . . . Grills . . . Hinges . . . Modular furniture . . . Picnic tables . . . Recreational vehicles . . . Servers . . . Vacation homes**

■ TAILGATING, once only an Ivy League custom, is now as much a part of the college and professional football scene across the county as the game itself. At West Point's Michie Stadium, where the photograph above was taken, tailgating Army rooters have on occasion displayed

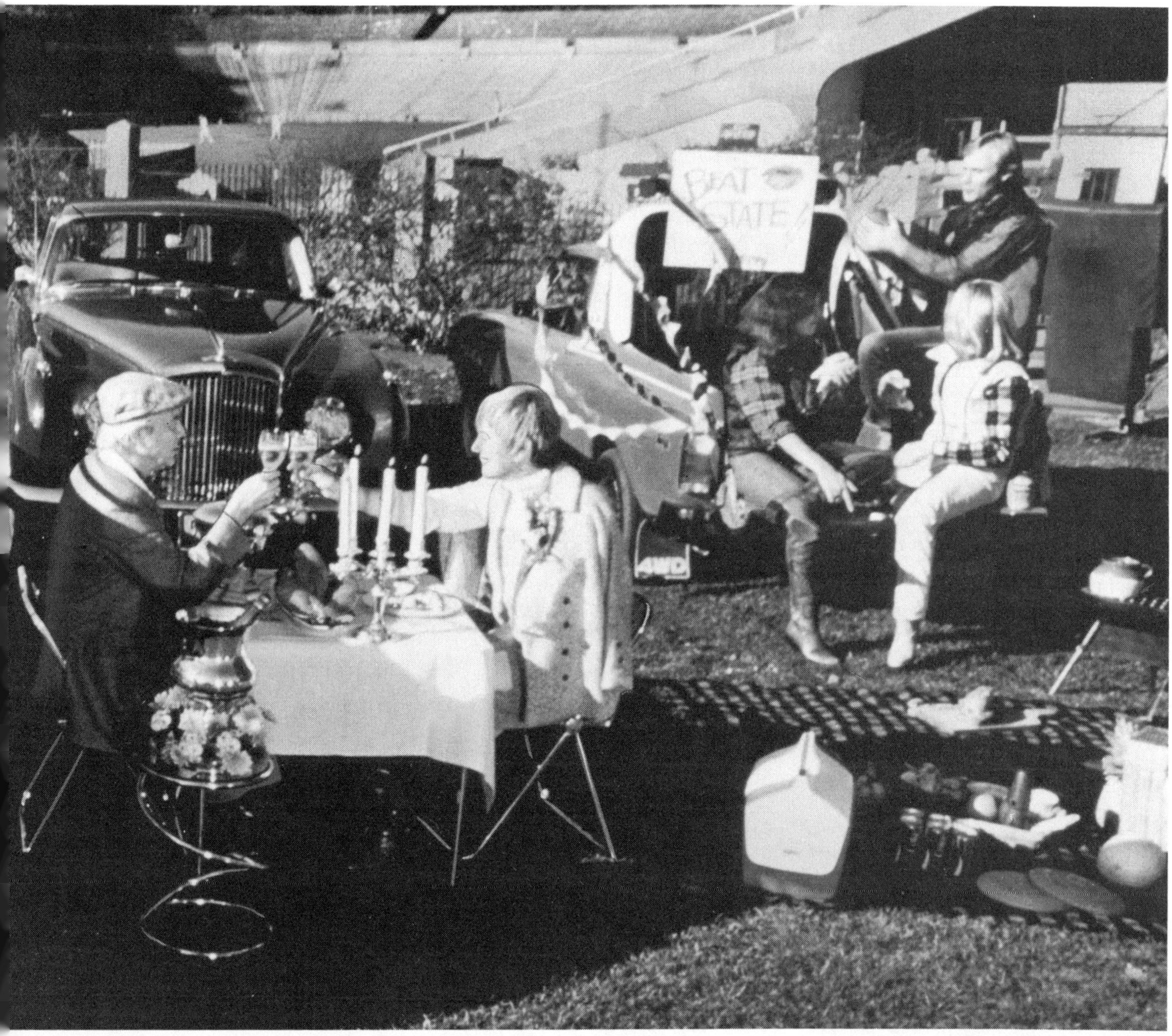


**HINGED DOOR** on left section opens to reveal a complete portable bar with mixing equipment. There is even storage for all popular bottle sizes.

**INSULATED** center section's drop-down door becomes service counter; has separate components for hot and cold foods, beverages, ice and stove.

**TABLEWARE** and utensils are secured in the right section. Each of the side sections can be separated from center section for individual use.

**CHEST IS** sized to fit most wagons; it is shown on a Datsun 810.

**AT HOME,** table-mounted or wall-mounted unit can be auxiliary food center.

**COMPLETELY** assemble the end cabinets. Neat-fitting doors are created by sawing box into two parts. Locate nails out of blade's path.

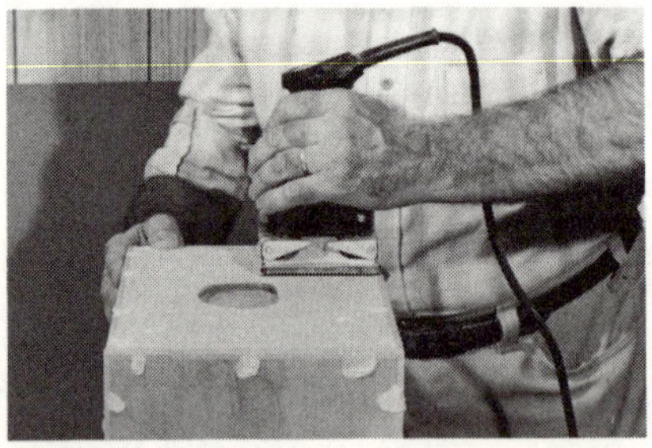

**SET ALL** nailheads, fill the holes and sand off all surfaces smooth before you begin to make the saw cuts that separate the doors.

## MATERIALS LIST —GENERAL CABINET

| Key | No. | Size and description |
|---|---|---|
| A | 2 | ¾ x 7½ x 17½"* |
| B | 2 | ¾ x 8⅝ x 18⅝"* |
| C | 4 | ¾ x 16⅜ x 17½* |
| D | 4 | ¾ x 8⅝ x 16⅜"* |
| E | 1 | ¾ x 17½ x 20½"* |
| F | 1 | ¾ x 18 x 21⅝"* |
| G | 2 | ¾ x 14 x 17½"* |
| H | 1 | ¾ x 14 x 21⅝"* |
| I | 1 | ¾ x 14 13/16 x 21⅝"* |
| J | 2 | ¾ x 18 x 42⅜"* |
| K | 1 | ¾ x 19½ x 42⅜"* |
| L | 2 | ¾ x 17¼ x 19½"* |
| M | 14 | ¼ x 1¼ x 42⅜" maple (rip from 6/4 stock) |
| N | 14 | ¼ x 1¼ x 21⅝" maple (rip from 6/4 stock) |
| O | (as reqd.) | 3/16 x ¾" maple strips (rip from 1" stock) |

*Birch veneer, lumber-core plywood was used in prototype.

Misc.: 10, ½"-dia. wood plugs; 26 No. 12 x 2" fh screws; 13, ⅜ x 3" lagscrews (for wall mounting of unit); 1" brads; 2" finishing nails and glue as required.

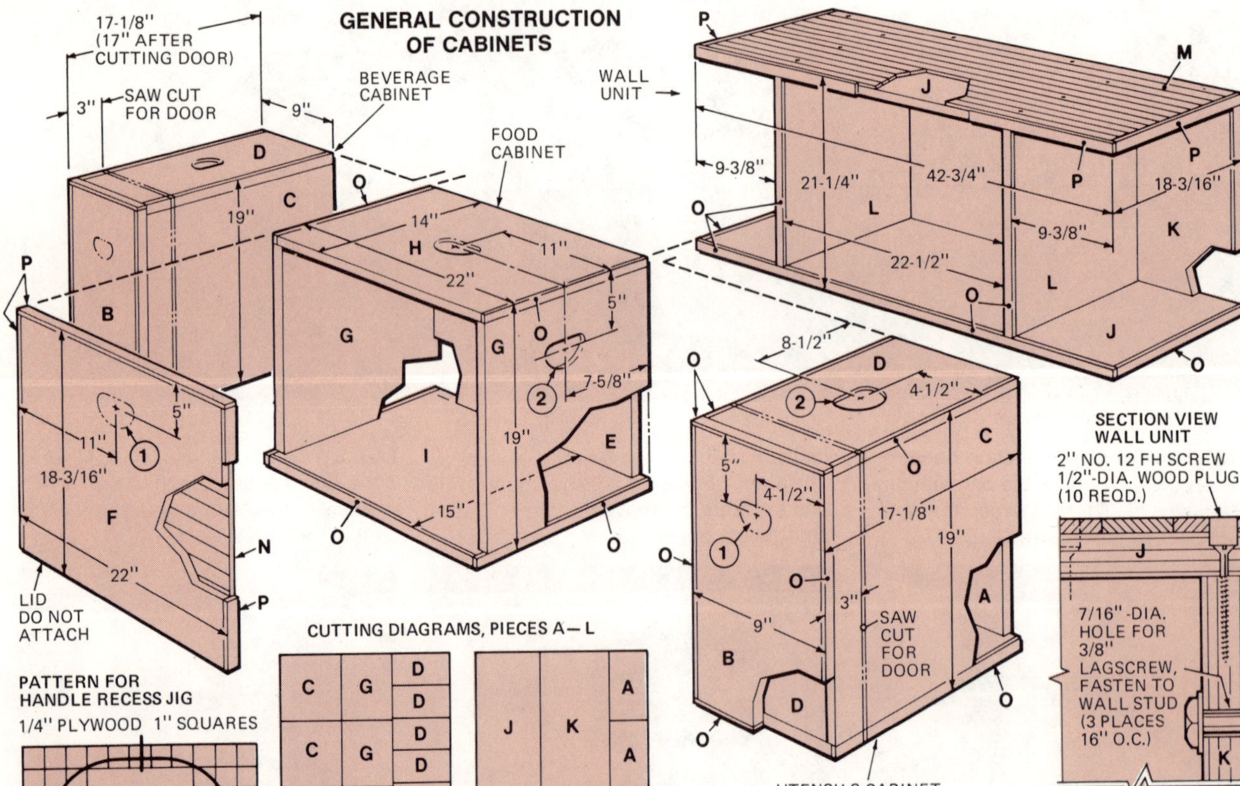



PATTERN FOR HANDLE RECESS JIG

1/4" PLYWOOD 1" SQUARES

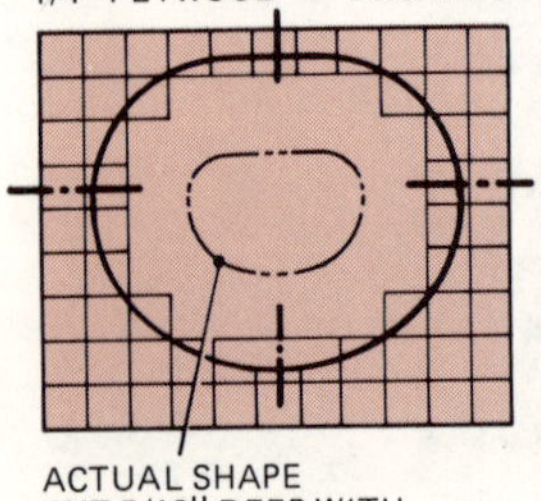



CUTTING DIAGRAMS, PIECES A—L

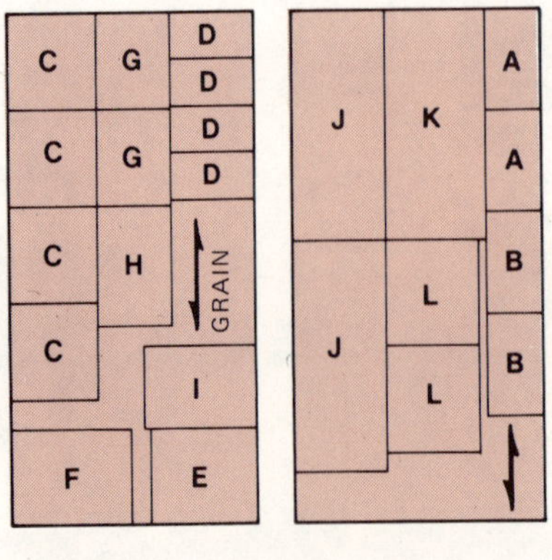



SECTION VIEW WALL UNIT

such status accessories as crystal wine service, electric rotisseries, a chandelier on a stand over linen napery—even a baby grand piano on a flatbed truck.

Everyone tries to outdo everyone else at these tailgating parties, but for outscoring the opposition where it counts—in the food and drink department—this great picnic chest is rated number one. It keeps all your supplies right at your fingertips and organizes the whole affair so you will be the envy of the sporting crowd.

At home you can park the chest in your family room. Use it as an auxiliary food service center when the guys come over for the Monday night game or as a portable beverage and snack bar at a party. You can be sure that all the members of the family will find a use for it when they do their entertaining.

**where every inch counts**

The chest makes use of just about every cubic inch for storage. The food cabinet has two separate, insulated compartments for storing perishables and/or keeping precooked dishes warm. Its front panel (F), with maple butcher block on one side, does double duty as a sandwich cutting board or as a serving tray.

Circular cutouts in the beverage cabinet keep bottles from bumping and breaking. Similarly, the tableware cabinet utilizes dividers to keep dishes from arriving at the stadium in pieces. Deep doors on the end cabinets are equipped with hooks and holders for mugs, utensils and bar equipment, plus a rack for napkins. These spaces can be rearranged to suit your own needs. You may want to begin by thinking through what you will need for a fully equipped outing and how you should pack it in.

**where to start**

You will need two 4 x 8 sheets of ¾-in. lumber-core plywood. We used birch cabinet-grade plywood for the prototype, but wood choice is optional. Since the plywood will cost approximately $100 (about half of what you'll spend on the entire project), carefully follow the cutting diagrams on page 496. You can get away with a lesser grade plywood and save yourself some money, but don't expect the quality of the finished product to be quite the same.

Lay out the cutting lines and tack-nail a straight board to the panel for guiding the saw while you make the cuts. Use a circular saw set to a depth so the blade will just cut through the ¾-in. plywood. Carefully plan your cuts so you can cut any long straight lines first. Be sure the saw kerf is on the waste side of the cutting line for each pass. When you reach intersecting lines, finish off the cut square with a sabre saw. Final accurate sizing can be done with the help of your table saw.

Edge strips (O and P) ripped from 4/4 southern maple stock are used to conceal plywood edges. Set the table-saw rip fence for a 3/16-in. cut and rip all the required strips from warp-free boards.

Use a few finishing nails to hold the edge strips in place when gluing them to the panel. Not all edges require strips, so check the drawings. Attach strips to the ends of *H, I* and *D,* to the front of I, to the ends and sides of B, to the fronts of L and to the sides and front of J. Leave strips off the edges of both cutting boards (F and J) until the maple strips (M and N) have been glued in place and the edges trimmed. The maple strips closely match the birch veneer and are easily worked with tools where required.

**handle recesses**

Recesses for the handles should be made before the boxes are assembled. For a quick, accurate job, use a template jig (for pattern, see plans on page 496) to guide the router and a ½-in. mortising bit. The router should have a 5-in.-dia. shoe. If yours is different, adjust template dimensions to suit. Use brads to secure the template to the workpiece.

Sand all exposed interior surfaces before assembling the cabinets. Use a pad sander or wrap sandpaper around a block and rub with a steady flat motion. Take care that you don't round the corners of the pieces that are to be butt-joined. This can result in sloppy looking joints.

Assemble the three cabinets using 2-in. finishing nails and a thin layer of glue at all butt joints. Start by attaching the sides to the back. Next, attach the tops and bottoms. Then add the fronts to the end units.

**TO SEPARATE** doors, set table-saw fence and cut through narrow ends first. Proceed to cut long sides. To prevent blade from scraping at end of final cut, insert cardboard into upper kerfs and use tape over the kerf.

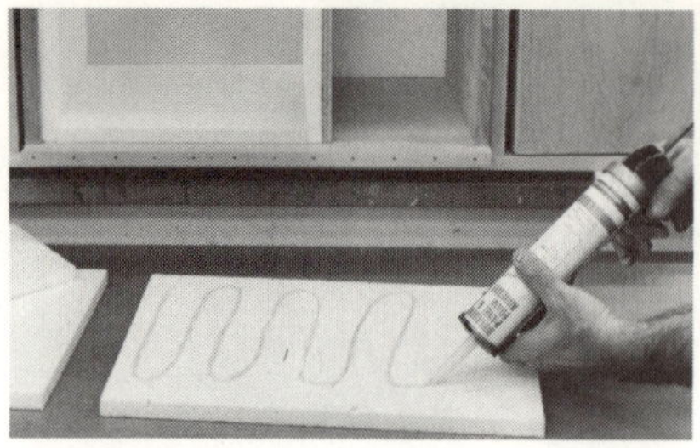

**APPLY GLUE** to polystyrene foam panel with gun-type applicator. Be sure adhesive is of the type formulated for foam. Paint exposed sides and edges of insulation with latex enamel to make them easier to clean.

**TEST-FIT DOWELS** in insulated compartments. Remove to paint. Dowels need no glue; press fit holds them.

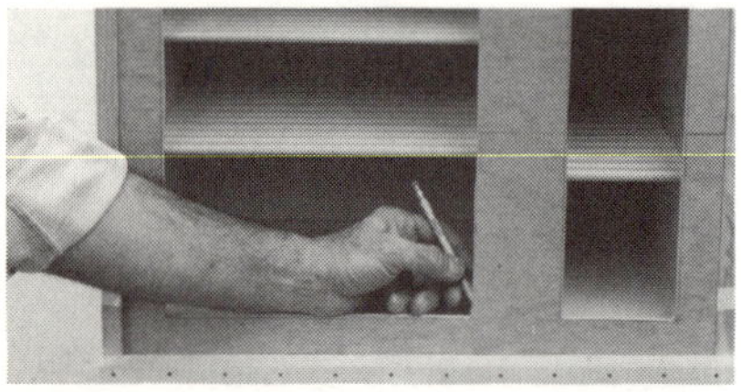

**AFTER** cutting rough undersize openings in front (C), trace exact openings using bottom (E) and sides (D) as guides.

**FINISH CUTS** on face framing are made with table saw. Set fence, lower blade. Then position work, turn on power, slowly elevate blade.

**INSULATING** panels, with edge strips in place, should be slightly larger than their openings to fit snugly. Sand the strips to exact size.

**MATERIALS LIST—FOOD CABINET**

| Key | No. | Size and description |
|---|---|---|
| A | 1 | ¼ x 11½ x 15⅝″ fir plywood |
| B | 1 | ¼ x 4⅞ x 15⅝″ fir plywood |
| C | 1 | ¼ x 17½ x 20½″ fir plywood |
| D | 4 | ¼ x 12 x 15 11/16″ fir plywood |
| E | 2 | ¼ x 12 x 12″ fir plywood |
| F | 1 | ¼ x 12 x 16″ fir plywood |
| G | 2 | ¼ x 5⅜ x 12″ fir plywood |
| H | 1 | ¼ x 5⅜ x 16″ fir plywood |
| I | 1 | ½ x 12⅞ x 17½″* fir plywood |
| J | (as reqd.) | ⅜ x 11/16″ maple (rip from 1″ stock) |
| K | 44 | ⅜″-dia. x 5¼″ dowel |
| L | 1 | 11/16 x 10¾ x 14⅞″ polystyrene rigid-panel insulation |
| M | 1 | 11/16 x 4⅛ x 14⅞″ polystyrene rigid-panel insulation |
| N | 1 | 11/16 x 13 15/16 x 17½″ polystyrene rigid-panel insulation |
| O | 2 | 11/16 x 12¼ x 17½″ polystyrene rigid-panel insulation |
| P | 2 | 11/16 x 12 x 12¼″ polystyrene rigid-panel insulation |
| Q | 1 | 11/16 x 6¾ x 17½″ polystyrene rigid-panel insulation |
| R | 2 | 11/16 x 12¼ x 17½″ polystyrene rigid-panel insulation |
| S | 2 | 11/16 x 5 5/16 x 12¼″ polystyrene rigid-panel insulation |
| T | 1 | Brass hasp (Brainerd No. 840XC) |
| U | 6 | ¾ x 1¾ x 1¾″ brass corners (Constantine No. 32H3A) |
| V | 4 | Plastic glide tack |
| W | 3 | Brass-plated handle (Brainerd No. 0234XC) |
| X | 4 | ¼ x 1″ rh bolt, washer and nut |
| Y | 1 | 1⅛ x 2″ solid-brass handle (Stanley CD5336N01) |
| Z | 1 | 1½″ x 21¾″ brass-plated continuous hinge |
| AA | 8 | ½″ escutcheon pin |

**Misc.:** polystyrene foam adhesive.

**FOOD-STORAGE CABINET, EXPLODED VIEW**

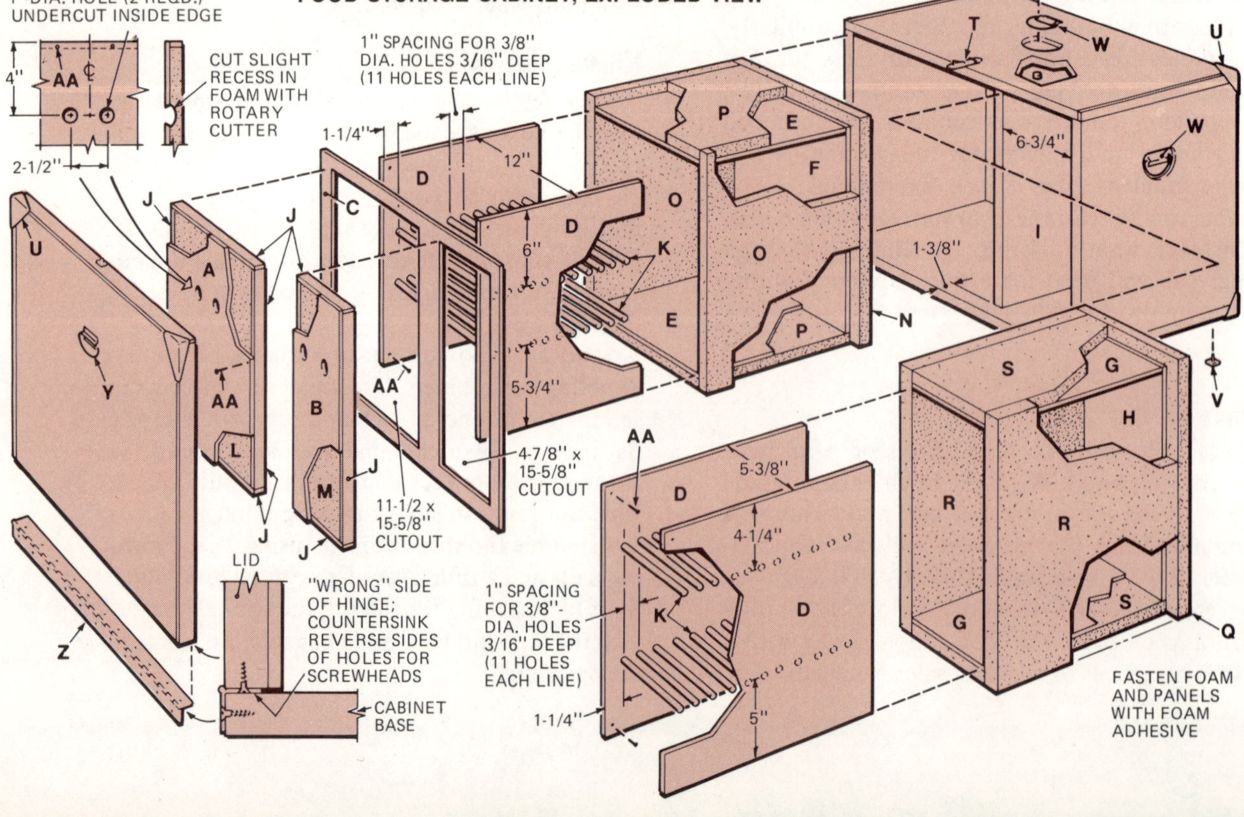

**perfect fitting doors**

The end cabinets are constructed as completely closed boxes which are then carefully ripped on the table saw to produce the 3-in.-deep doors. This method results in matched grain and precise-fitting doors. When nailing the units together, be sure to keep the nails out of the anticipated cutting lines. Some extra planning here will save trouble later.

To separate the doors from the end cabinet boxes, set the saw fence at 14 in., with the door portion as the cutoff. Working with one box at a time, cut through the smaller end sections first. Make the third cut on one of the sides. Before making the final cut, insert and tape cardboard spacers (the exact width of the kerf) into each end of the kerf on the first side. Also, apply several pieces of masking tape over other portions of the kerf. This prevents the cutoff section from accidently pressing against the side of the blade at the completion of the cut.

**cutting polystyrene panels**

Polystyrene rigid insulating panels are used to insulate the center unit. The best way to cut the foam is with the table saw. The blade you use *must have set teeth.* Do not use a hollow-ground blade—it'll positively bind and could cause an accident. Note: Because of chest weight, the top handles for the chest are attached with 1-in. rh nuts and machine screws, instead of with the screws supplied with the handles. As a result, the handle for the center chest must be installed *before* the insulation goes in. This allows you to use a wrench on the nuts.

The dowel shelf assemblies should also be completed before installing the insulation. The best way to insure exact alignment of the dowel holes between side pieces (D) is to tape the pieces together and drill 1/16-in. holes through both pieces at each dowel location. Then, using a brad point drill bit (which bores a flat-bottomed hole), counterbore holes 3/16 in. deep x 3/8 in. in dia.

Press the sides onto the dowels (glue is not necessary). Spray-paint the shelf assemblies before proceeding.

**plywood 'skin' for foam**

When you're ready to make the installation, apply foam panel adhesive to the backs of the insulating panels and set them in place. The foam panels are not made to take abuse, so a 1/4-in.-plywood protective skin must be applied over them. Install pieces E, F, G and H first using the adhesive. Then install the shelf assemblies; since the fit is snug, they do not need gluing.

The front frame piece (C) on the center section can be made to fit perfectly in only one way—by marking the cutting lines from the inside with the piece in place. Make a rough cut to clear out most of the waste. This will enable you to get a pencil on the inside to trace exactly along the sides of the enclosure.

**applying butcher block surfaces**

To make the butcher block surfaces, cut the panels (F and J) slightly long so the ends can be trimmed to size after the maple strips (M and N) have been glued on. Use masking tape to hold all the strips together. Plastic resin glue is the best for this job. It has a water-resistant quality that will hold the strips; they will likely become damp during use. Mix it slightly thinner than usual so it won't set up too fast. Spread it over both surfaces and make sure to get it on the edges between the strips.

The wall unit is put together with screws and glue. Counterbored holes are made in the top to allow the screws to be recessed and plugged. Use bar clamps to hold the parts together while boring pilot holes for the screws. To assemble, connect the uprights to the back, add the bottom and then the top. See the plans for fitting out the insides of the end cabinets. Install all remaining hardware.

**finishing touches**

The insides of the prototype chest were painted a yellow enamel. To achieve the light tone on the chest's exterior surfaces, they received a wash coat of 1 part sanding sealer to 8 parts thinner before being stained with Benjamin Moore's Fruitwood stain. Finally, two coats of a satin-finish polyurethane varnish were applied. Carefully sand between coats to obtain the best finish. Do not stain the cutting boards. Use a finish specially formulated for surfaces that contact food, such as Constantine's Wood Bowl Seal, or rub with mineral oil. If you decide to rub it with mineral oil, periodically rub some more into the surface to maintain a beautiful and durable finish.

Your tailgater's picnic cabinet should now be ready to use. Get the gang together, pack the supplies and enjoy the party—and the game.

*tailgate cabinet, continued*

## MATERIALS LIST— BEVERAGE AND UTENSIL CABINETS

| Key | No. | Size and description |
|---|---|---|
| A | 2 | ¼ x 7½ x 13⅛″ fir plywood |
| B | 1 | ¼ x 7½ x 13½″ fir plywood |
| C | 2 | ⅜ x 3⅞ x 12⅝″ fir plywood |
| D | 2 | ⅜ x 3⅞ x 7½″ fir plywood |
| E | 1 | ⅜ x 3⅞ x 3″ fir plywood |
| F | 2 | ½ x ¾ x 13½″ maple |
| G | 1 | ½ x ¾ x 7″ maple |
| H | 2 | ⅜″-dia. x 5″ dowel |
| I | 2 | 1⅛ x 2″ No. 1 solid-brass chest handle (Stanley CD5336) |
| J | 2 | brass-plated handle (Brainerd No. 0234XC) |
| K | 16 | 1¾ x ¾ x ¾″ brass corner (Constantine No. 32H3A) |
| L | 2 | hasp (Brainerd No. 840XC) |
| M | 2 | 1¹⁄₁₆ x 19¾″ brass-plated continuous hinge |
| N | 8 | plastic glide tack (do not install on doors) |
| O | 3 | brass cup hook |
| P | 2 | ¼ x 7½ x 12⅛″ fir plywood |
| Q | 2 | ¼ x 1¼ x 12⅛″ fir plywood |
| R | 1 | ¼ x 1½ x 5¼″ fir plywood |
| S | 1 | ¼ x 1¾″ x 5¼″ fir plywood |
| T | 2 | ¼ x 1½ x 1½″ fir plywood |
| U | 1 | ¼ x 1¾ x 17½″ fir plywood |
| V | 1 | ¼ x 1¾ x 3¼″ fir plywood |
| W | 1 | ¼ x 1½ x 3¼″ fir plywood |
| X | 1 | ¼ x 1¾ x 3¾″ fir plywood |
| Y | 1 | ¼ x 1½ x 3¾″ fir plywood |
| Z | 4 | ¾ x 1½ x 1⅜″ maple |
| AA | 4 | ⅜″-dia. x 5⅞″ dowel |
| BB | 2 | ⅜″-dia. x 5⅞″ dowel |

**Misc.:** 4 No. 8 x 1¾″ fh screws; 1″ rh machine screw, washer and 4 nuts (for handle J); 1″ brads, ½″ escutcheon pins, white glue and 2″ finishing nails.

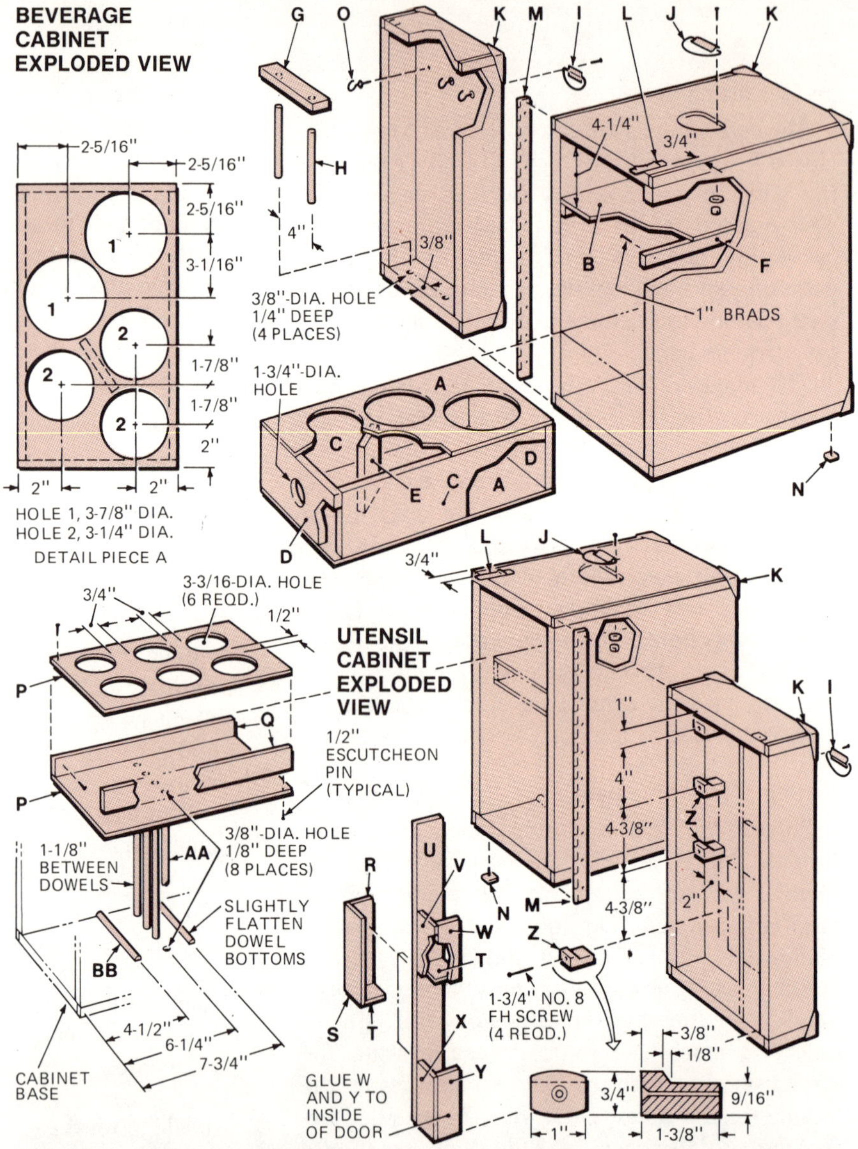

**TO ASSEMBLE** cabinet top with strips (M), use tape on good side. Then flop piece and spread strips (tape acts as a hinge) to apply glue between.

**STRIP ASSEMBLY** is then glued to plywood (J). Use wax paper between clamping blocks and the work surface to prevent accidental bonding.

**USE BAR CLAMPS** to hold dividers (L) as shown when you bore pilot holes for screws. Be sure dividers are located correctly and are square.

**BORE HOLES** for plugs (to hide screwheads) *before* boring pilot holes.

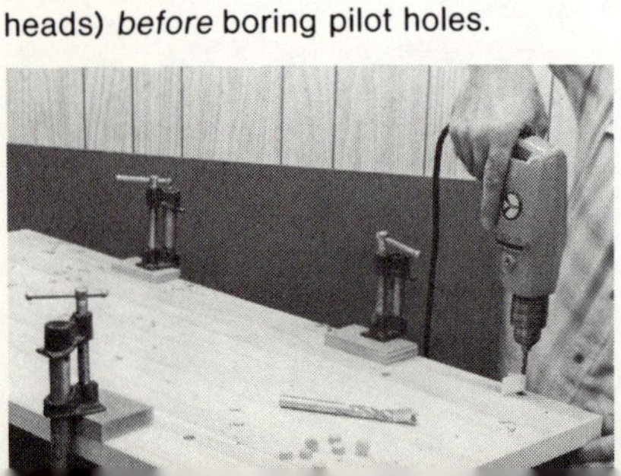

**USE BRUSH** to apply small amount of glue in holes for plugs.

**SUPPORTING UNIT** can be mounted on wall as shown in the plans.

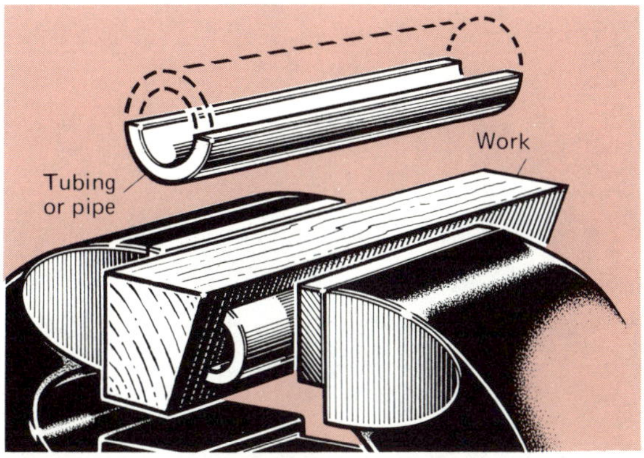


**IF YOU EVER** had to clamp a workpiece that has tapered sides in a vise, you know that it is difficult, because the piece keeps "squeezing" out. I've found that an easy way to keep it from moving is to place a half section of tubing or pipe between the tapered face and vise jaw. This gives a rocking action which insures tight clamping.—*Harvey Muller*

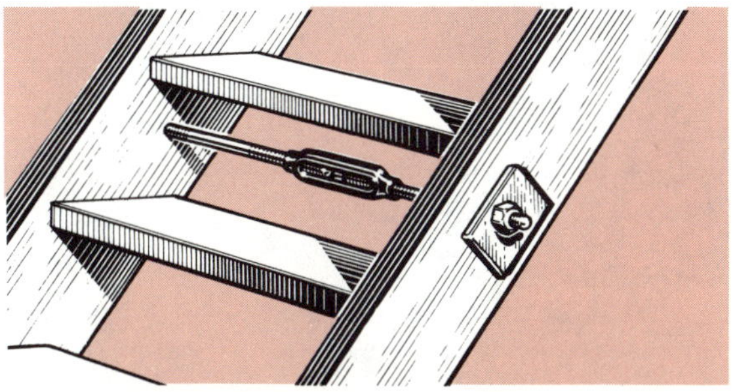

**BASEMENT STAIRS** are usually of simple construction with treads let into dadoes in the stringers. Occasionally, a few treads will start to leave their grooves. To pull the stairs tight again, try using a turnbuckle through the stringers as shown. Use a hefty-size steel plate to serve as a washer to insure against fracturing or splintering the wood.—*George Mayerchak*

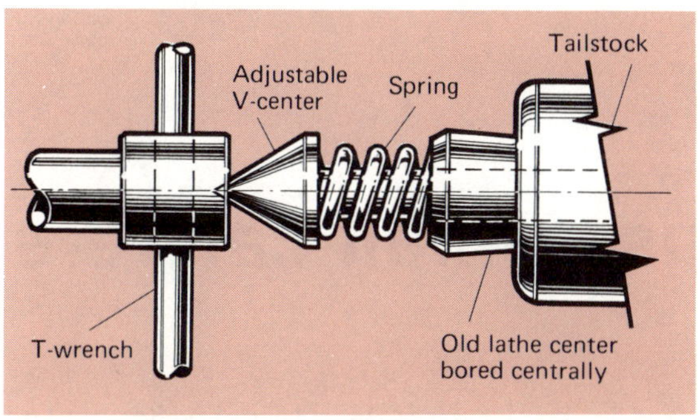


**HERE'S AN AUXILIARY** tailstock center that is self-compensating when it's used to guide a tap wrench to thread chuck-held work. It's spring-loaded, so it can't exert undue pressure on the tap as it is being fed into the work. It's made by center-drilling an old lathe center, turning a V-point to slip inside and adding a light compression spring.—*R. J. Hill*

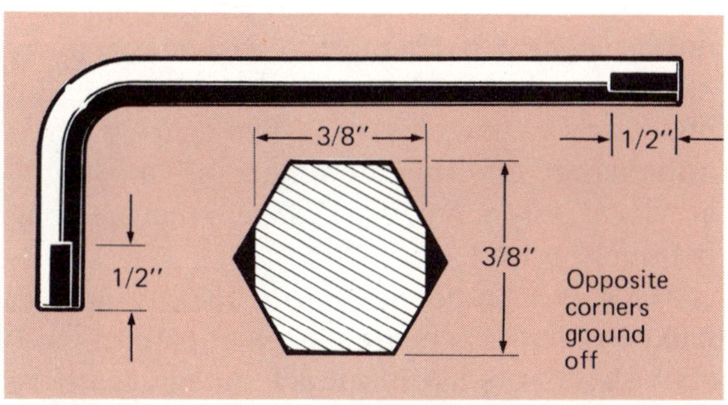


**YOU CAN MAKE** a good long-and-short handle for your socket-head wrench, and still allow the hex wrench to perform its intended function. First, select a hex wrench with an over-the-flats dimension of your square-drive socket wrench. Then grind off two opposite corners of the hex cross section for about ½ in. at the ends.—*Anton M. Barkey*

**FOR YEARS** I have used the method shown to cut all sizes of bolts and nails without damaging threads or nailheads. And the best part is, it saves looking for scrap to drill out and hold the piece for cutting. Simply mount the workpiece in either type of clothespin and then clamp the entire set-up in the vise. Try it; it works perfectly.—*A. J. Porder*

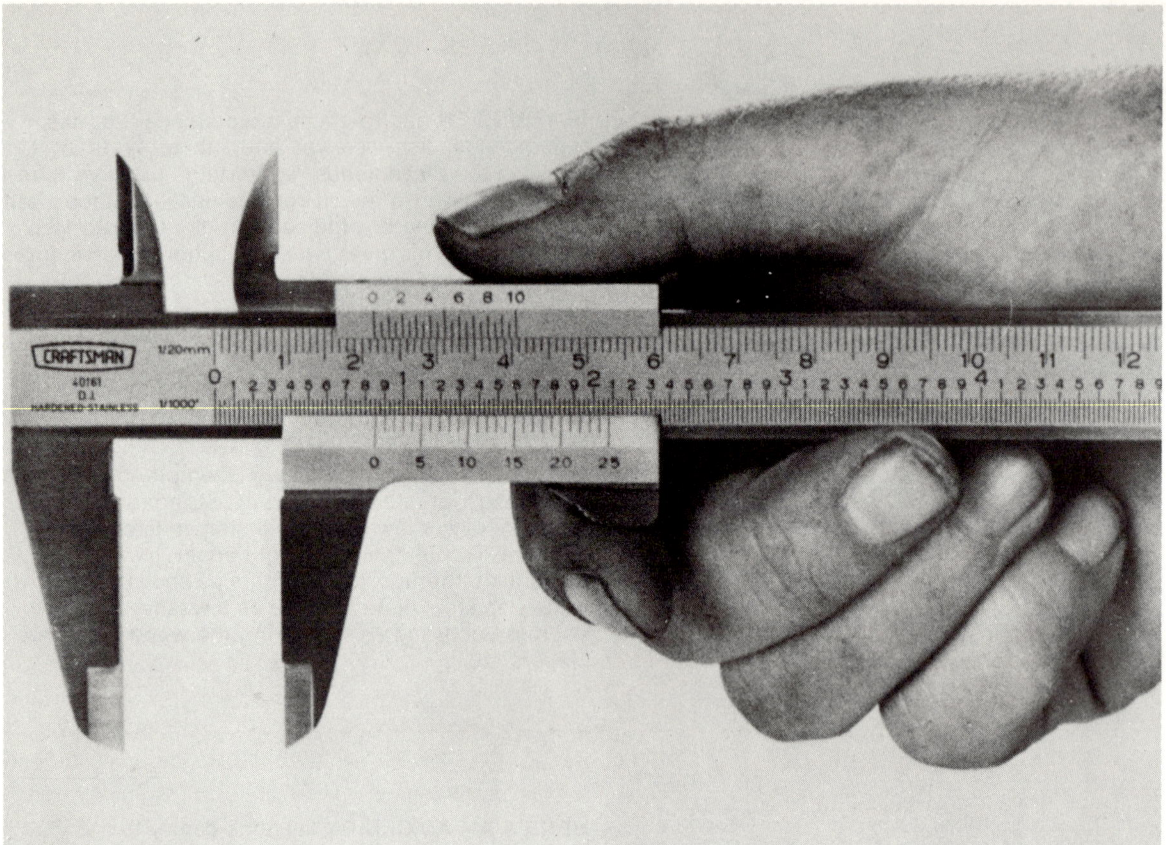


# How to choose and use small calipers

By JOHN BURROUGHS

**Now, in home workshop and metalworking, you can employ a precision measuring tool that has long been a standby of machinists. Formerly too expensive for home use, vernier calipers with gnat's-whisker accuracy are now available at relatively modest prices**

SEE ALSO

**Measurements . . . Metric system . . . Micrometers . . . Sheet metal . . . Shop techniques . . . Tools, hand**

ALTHOUGH MACHINISTS have used them for years, precision vernier calipers have until recently been far too expensive to be known as popular home workshop tools. But now there's a new class of vernier calipers that is appearing on the commercial market—serviceable 6-in. precision calipers that are competitively priced.

Made in Europe and Japan, these relatively inexpensive calipers are universal pattern models with scissors-like inside jaws integral with the outside jaws. They are beautifully finished instruments—hardened stainless steel throughout, with nicely fitted slides, accurately lapped jaws and legible satin-chromed photoengraved scales.

The photoengraved scales account for the tools' relatively low prices. Best-quality 6-in. vernier calipers have machine-divided scales engraved on pantograph engraving machines. Photoengraved scales are made by a process similar to that used to make printing plates; that is, they're printed from glass master negatives and then etched. While the etched graduations aren't quite as crisp as the machine-cut graduations, you'd need a magnifying glass to tell the difference. Indeed, for general shopwork a photoengraved caliper is completely satisfactory.

Actually it's about the handiest pocketsized measuring tool imaginable for home workshop wood and metalworking. When you want a precision measurement, you can use the caliper's vernier to get a reading within a thousandth of an inch. When gnat-whisker accuracy isn't really required—when you're measuring the thickness of lumber, for example—you can forget the vernier and use the tool as a plain slide caliper.

A precision reading takes a little mental arithmetic, but this soon becomes second nature. All verniers are alike in principle. They're simply auxiliary scales used to subdivide the smallest graduations on the caliper bar's scale into as many parts as there are graduations on the vernier. Thus, you could consider a vernier as just a system of offsets.

### how to use the vernier

*The 25-division verniers used with the decimal-inch scales are 25 thousandths of an inch shorter in total length than the equivalent graduations on the bar scale.* This means that

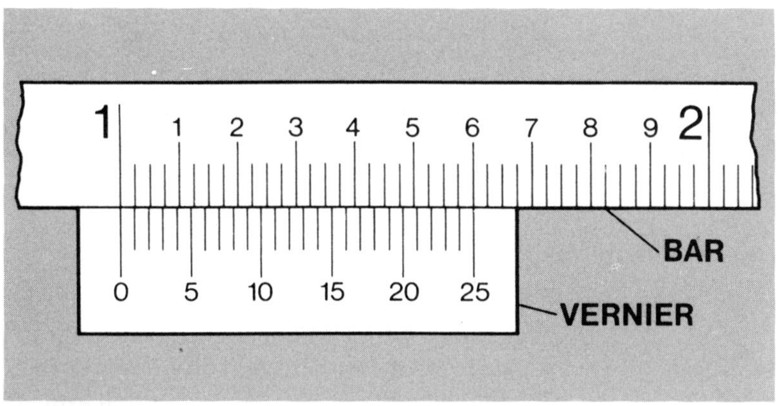


**SHORT** 25-division scale often found on older types of calipers is seldom used on newer models. The bar's scale is divided into tenths of inches, with each tenth divided into four parts or fortieths. The vernier subdivides these 1/40-in. graduations into 25 parts, each part being a thousandth.

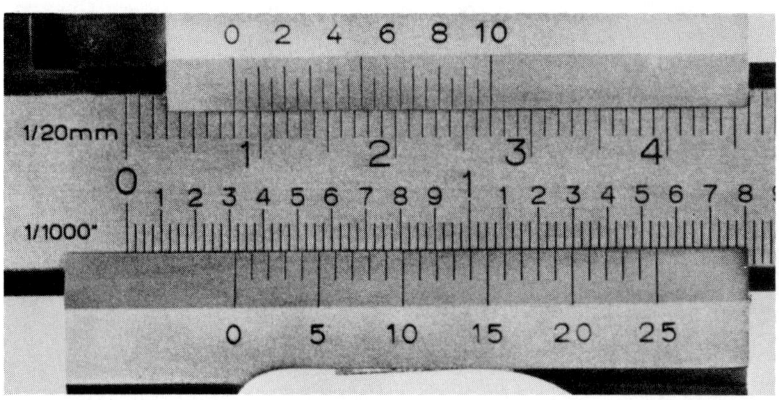


**LONG** 25-division vernier found on most of the new calipers is similar to short vernier, but the vernier graduations are double-spaced to make it easier to take accurate measurements. On the thousandths vernier, the reading is .317 in. The top vernier, indicating millimeters, divides metric scale's millimeter into twenty parts. Can you make the equivalent reading in millimeters?

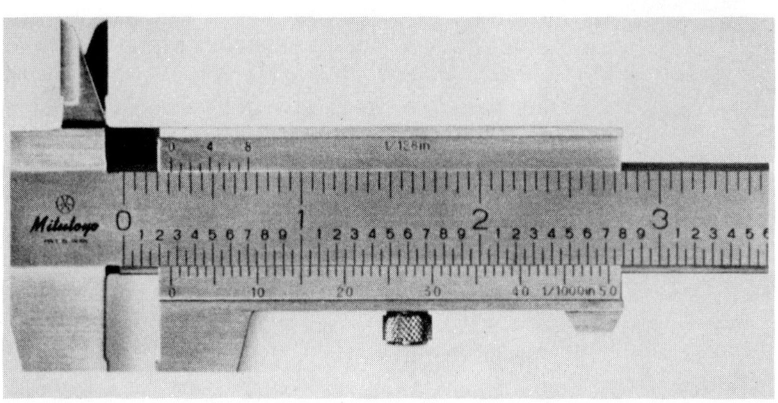


**EXTRA-LONG** 50-division vernier is featured on some newer makes of calipers. The bar scale has numbered tenths, each of which is divided into two parts, or twentieths. The vernier subdivides these 1/20th graduations into 50 parts, or thousandths. The reading shown is .265 inches. The top vernier scale divides the fractional-inch scale divisions (1/16ths) into eight parts or 1/128ths. What is equivalent fractional reading?

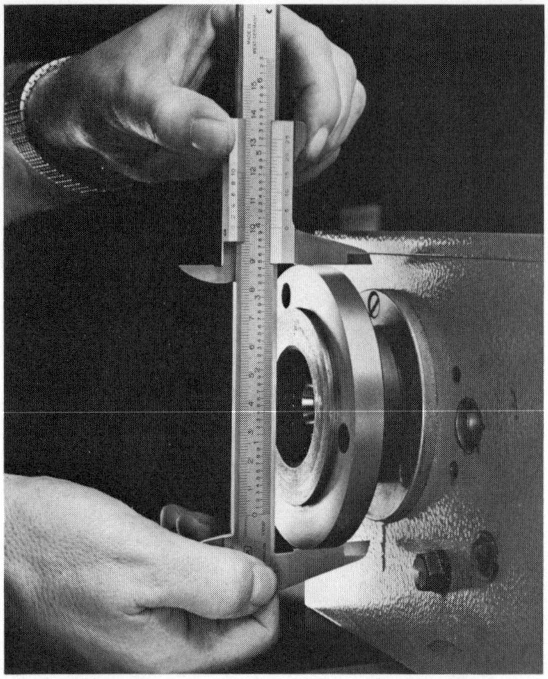

**CALIPERS ORDINARILY** are accurate to within one thousandth, more than sufficient for most work.

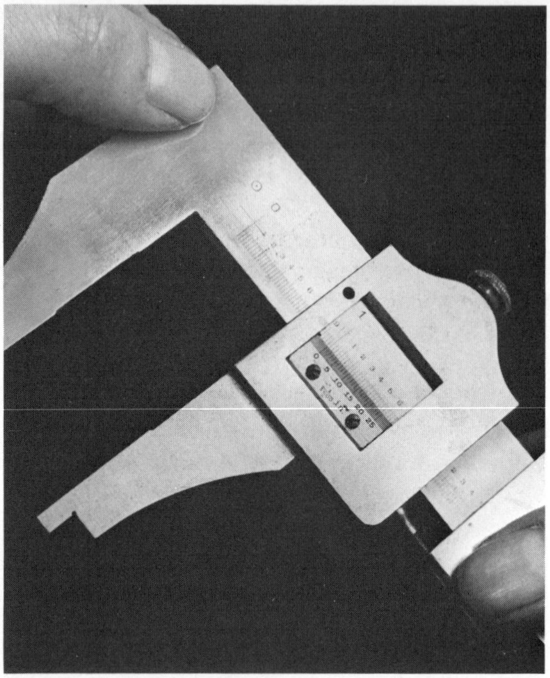

**OLDER STYLE CALIPERS** have 25-division verniers only .6 inches long, require use of a magnifying glass.

**TO MEASURE DIAMETERS** hold caliper square with work, take more than one reading for minimal error.

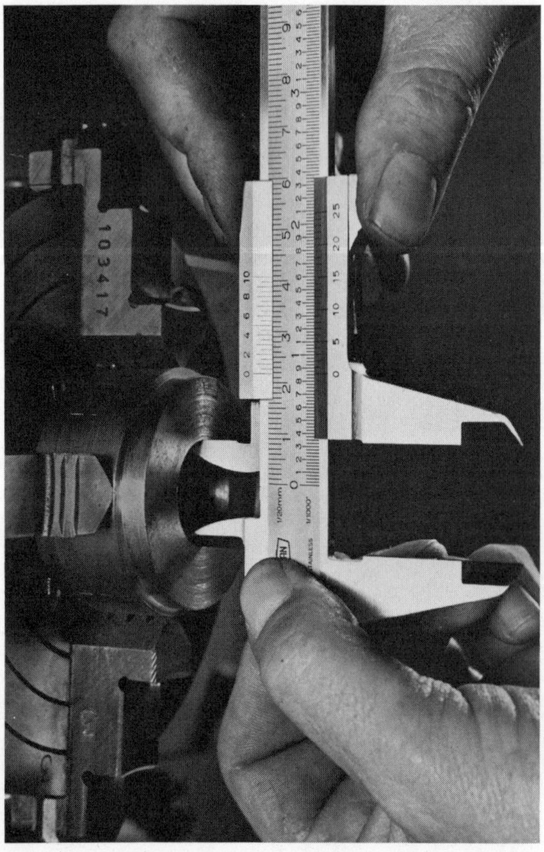

with the caliper closed, the vernier's zero-line is aligned with the bar's zero-line, but the vernier's first graduation is offset one thousandth from its corresponding line on the scale above; the second graduation is offset two thousandths; the third three thousandths; and so on.

With the caliper opened, the vernier zero-line indicates the measurement to be read. Thus, to read the bar scale, first note the number of whole inches (to the left of this line), then add the tenths of an inch, which are numbered. Next, count the smallest graduations, which are fortieths of an inch, or 25 thousandths each, and add them. Finally, using the vernier, measure and add any fraction of a small graduation remaining.

*Measuring this extra fraction* is simply a matter of running your eye along the vernier scale to see which graduation matches a graduation of the bar scale. For example, if the vernier's 17th line matches a line above, the vernier's zero-line is offset 17/25ths of a graduation, or 17 thousandths past the last graduation on the bar scale.

Simply stated, if you think in terms of "whole inches plus 100-thousandths plus 25 thousandths plus thousandths," it's easy to quickly make any reading in thousandths mentally.

**special features**

Universal-pattern vernier calipers have two

**DEPTHS CAN BE MEASURED** with the sliding jaw or rod on end of caliper beam. Vernier indicates reading.

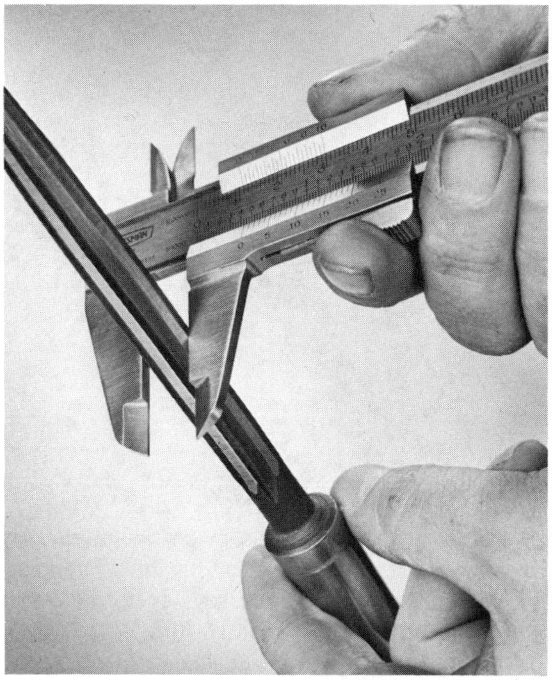

**USE SPECIAL CARE** when measuring hardened-steel cutting tools to prevent damage to caliper's jaws.

bar scales, upper and lower. The decimal-inch scale (inches graduated in tenths) is the more frequently used. The other scale may be either metric (centimeters and millimeters) or ordinary fractional-inch (inches graduated in sixteenths).

A metric scale's vernier subdivides the bar scale's millimeters into twentieths or fiftieths. A fractional-inch scale's vernier subdivides the bar scale's sixteenths into eight parts, or 128ths of an inch. On most calipers the two scales have a common zero-line, which permits reading equivalents—thousandths of an inch to millimeters, for example.

The slides of some 6-in. models have thumbscrew locks, while others have thumb-released friction locks. Some tools have fixed vernier plates—others have plates that can be adjusted to compensate for jaw wear. There are models with carbide-tipped jaws or fine-adjusting microscrews.

While a micro-screw caliper is considered preferable for machinework, a tool with a friction lock is more convenient for home workshop use.

No matter which model you choose, preserve your caliper's precision by taking care not to mar or spring its jaws. Also wipe the bar often with a lightly oiled rag to prevent grit or metal chips from working under the slide.

**CALIPER JAWS** are lapped until exactly parallel, so they can be used to check if work planes are parallel.

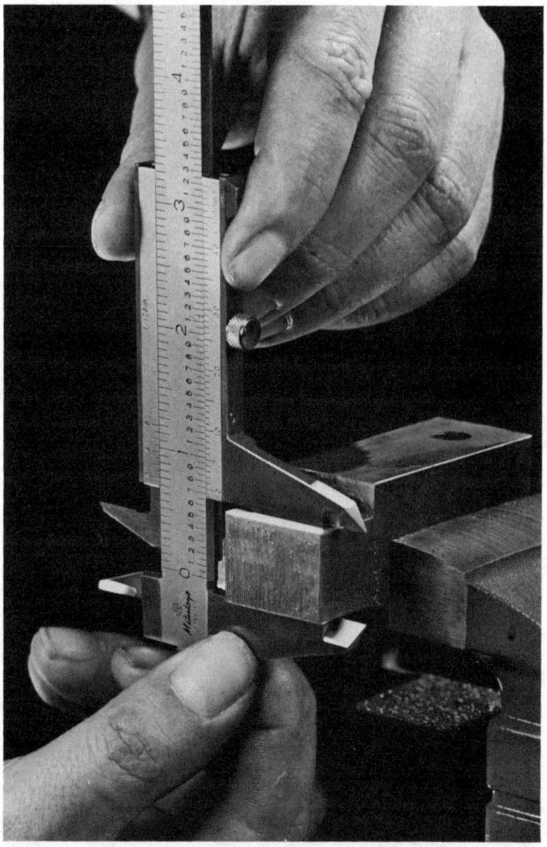

# What you should know about underhood cables and wiring

By MORT SCHULTZ

■ ONE OF THE DEMANDS made by battery cables and primary wiring, in particular, is that their connections be clean and tight. If connections aren't clean and tight, resistance to current flow will be high, and hard starting and stalling are likely to result.

### cause of sudden stalls

Primary wiring has been the cause of one of the most baffling failures experienced by car owners. Suppose you are rolling blissfully down the road when—pfft—just like that, the engine stalls. You brake to a halt, turn off the ignition key, scratch your head, turn on the key and crank the engine.

It starts, and the problem doesn't occur again until some later time. What may be causing it?

A loose primary connection could be the reason. Or maybe primary wire insulation has worn down and bare wire is shorting against the distributor housing.

Primary wiring must be kept in good condition. If it shorts out, there will be no voltage, no ignition and no go. Would you think to look inside the distributor for a bare wire?

### sparkplug cables

Obviously, sparkplug cables extend from the distributor cap to the sparkplugs. However, included in this designation is the single high-tension cable that runs from the tower of the ignition coil to the center tower of the distributor cap. Don't forget about it. When it fails, your engine won't start.

Cables don't last forever. They wear out as they get older because they are exposed to high underhood temperature, oil splash and moisture. These cause insulation to become brittle and crack, which gives electricity an area of least resistance to penetrate. Electricity will arc through cracks in the insulation and, accordingly, will be diverted from the sparkplugs.

Reduced current leads to a decrease in combustion efficiency and an increase in fuel consumption. When a sparkplug fails to ignite the fuel mixture, fuel in the affected cylinder is thrown out the exhaust. It is wasted. This is what's known as misfire.

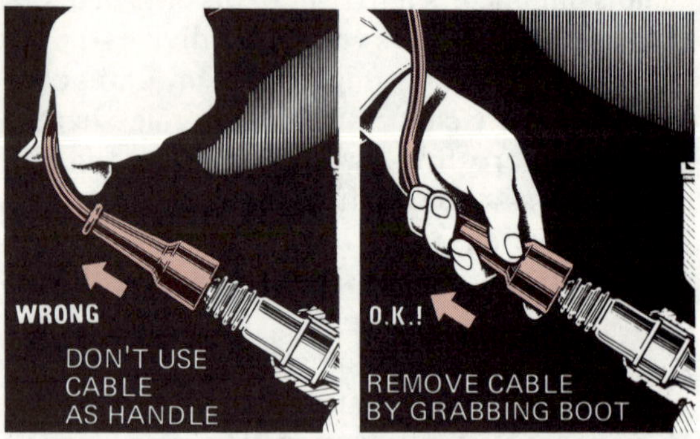


**TREAT** your sparkplug cables gently and they won't let you down. Don't pull on the cable; use a special pliers or grasp the boot firmly.

Environment isn't the only factor leading to cable failure. Mechanics employing bad work habits are another. Cables should not be tugged on when they are being disconnected. Strain causes conductive strands to separate, which leads to excessive resistance to current flow. Again, this will result in engine misfire.

SEE ALSO
**Auto batteries . . . Battery testers . . .**
**Dieseling, auto . . . Distributors, auto . . .**
**Electronic ignition, auto . . . Engines, auto . . .**
**Ignition systems, auto . . . Sparkplugs . . .**
**Tune-up, auto . . . Winterizing, autos**

**handling cables properly**

To handle cables properly, use a pair of sparkplug cable pliers. The tool costs a few dollars, but is worth the money. It helps prevent accidents that can cause cable damage.

To remove cables from sparkplugs using these pliers, grasp the boot covering the plug with the tool, and twist back and forth as you pull the boot from the sparkplug.

If sparkplug cable pliers aren't available, remove cables by firmly grasping the boot and twisting back and forth as you pull. Do not pull the cables. Pull only on the boots!

Cables have two ends. It's just as important to treat them properly at the distributor end as it is at sparkplugs. To remove a cable at the distributor, grasp the boot over the tower. Twist it back and forth as you pull. Do not pull the cables. Pull only on the boots!

**testing cables**

Test cables when you do an engine tuneup. Testing is done in two ways: by inspection and with an ohmmeter.

Look for cracks in insulation. And don't forget the boots. A boot is part of a cable and tends to dry out and crack, too. A bad boot will cause flashover: Electricity will take the path of least resistance and come out through a crack in the boot, sparking across the outside of the sparkplug insulator. Electricity won't get to where it's needed—to the electrode end of the sparkplug.

Unfortunately, visual inspection isn't a foolproof test. That is why I like the probing method, which will reveal a bad cable when failure isn't apparent to the eye.

You need a good-size screwdriver and a jumper wire that has an alligator clip on each end. Fasten the alligator clip on one end to the shank of the screwdriver and fasten the other end to a clean ground on the engine.

Start the engine. Remove a cable from a sparkplug. Probe the disconnected cable and its boot with the tip of the screwdriver. A spark that jumps from the cable to the screwdriver reveals an area of weakness.

**current leak**

After testing all cables, leave any one sparkplug cable disconnected as you probe the cable and boots between the coil and the center tower of the distributor. One sparkplug cable must remain disconnected while you check the coil-to-distributor cable so that one of the paths of least resistance which the ignition spark would normally follow is removed. Thus if there *is* a current leak in the coil-to-distributor cable it will jump to the screwdriver once during each revolution of the distributor rotor.

*Caution:* If your car is equipped with a catalytic converter, don't keep sparkplug cables disconnected for any longer than 30 seconds at a time to avoid catalytic converter overheating. Test as many cables as you can in 30 seconds and then test the others after the converter cools down.

If one or more cables are found defective, replace them all. If one bad cable is uncovered, chances are that the others will soon peak out.

When checking sparkplug cables, it is important that you inspect the metal connectors inside the boots on both ends. Corroded or burned connectors will cause excessive resistance. Replace the cables if this damage exists.

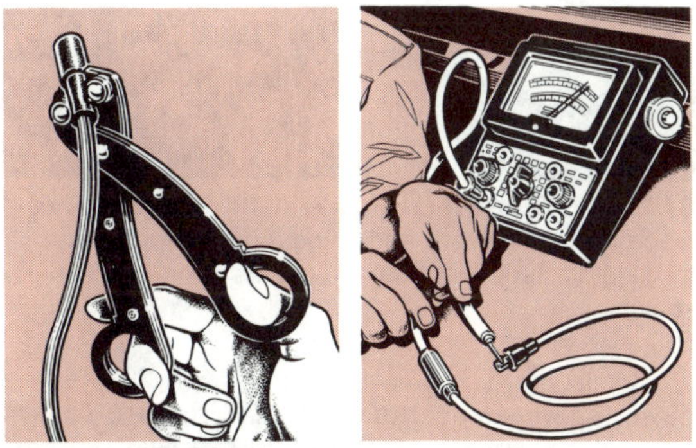

**SPARKPLUG** pliers (left) are the best way to pull boots. Use an ohmmeter (right) when you measure the resistance of your cables and other wires.

**secure cable connections**

*Important:* The main reason for a burned connector (it will be black) is failure to connect cables securely. Keep this in mind whenever you are reconnecting a cable to its sparkplug or seat in the distributor.

After examining cables, test them for excessive resistance to electricity with an ohmmeter that has a scale which reads in thousands of ohms. Lay the cable on a workbench. Turn the ohmmeter on and connect its leads to each end of the cable.

The amount of allowable resistance depends on the length of the cable and may even be marked on the cable. Generally, however, if resistance is greater than 25,000 ohms, discard the cable.

After testing one cable, reinstall it before removing another. This will prevent you from connecting a sparkplug to a wrong tower of the distributor, which will result in out-of-turn plug firing and possible engine damage.

When you reconnect cables, be sure they don't rest against a part that can cut or burn them, such as the throttle linkage or exhaust manifold. See that cables crisscross one another to prevent crossfire.

**crossing up crossfire**

Crossfire, which is also called induction leakage, occurs when sparkplug cables are grouped

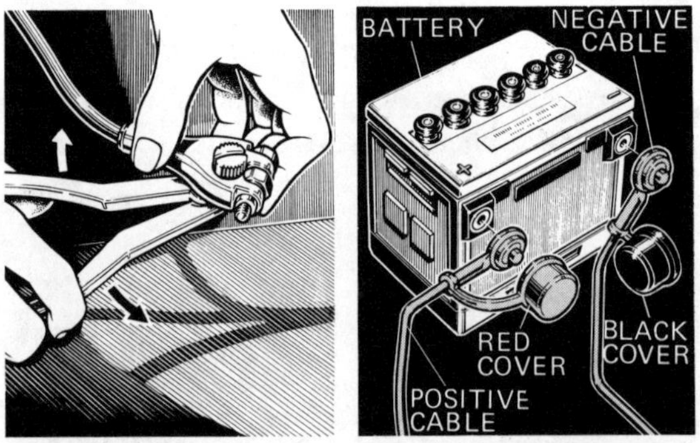


**BATTERY TERMINAL** spreader (left) is needed to replace terminals easily and without hammering. Side terminal batteries (right) have cables connected as shown.

close together and run parallel. Crossfire causes engine roughness, backfiring and detonation. Engine and sparkplug damage may result.

Crossfire refers to a sparkplug firing out of time because of induction. This means that the strong magnetic electrical field that envelops a sparkplug cable will be absorbed by an adjacent sparkplug to fire out of turn. The condition prevails between two *consecutively* firing cylinders.

The best way to avoid crossfire is to make sure that cables crisscross as they approach sparkplugs so no two physically adjacent cables run parallel. Most automobile manufacturers equip cars with engine-mounted looms that hold cables and help maintain a proper arrangement. If a loom is damaged or lost, replace it.

**battery cables**

What you should know about battery cables can be summarized as follows:

1. The cable parts of battery cables don't cause problems nearly as often as the terminal parts of battery cables, especially at battery posts.

2. The best way to prevent trouble from battery-cable terminals is to service them semiannually to prevent corrosion that interferes with electrical flow. Take cables off the battery posts. Useful tools you should consider purchasing if battery posts are on top of your battery are the following:

■ Terminal nut wrench and battery pliers to loosen the terminals' bolts and nuts. Hold the bolt stationary with the wrench and turn the nut with the pliers. Always remove the negative cable first and connect it last, because the path to ground is broken and thus the shock hazard is reduced.

■ A terminal puller whose jaws engage the bottom of the terminal so when the tool's screw is tightened, the terminal is drawn evenly off the battery post.

*Caution:* Don't bang on the terminal with a hammer to loosen it, and don't try prying it off with a screw driver. That's an excellent way to ruin the battery.

■ A battery cleaning tool to polish terminals and posts. Wash terminals and battery posts with a baking soda solution. Then polish the insides of the terminals with the terminal brush end of the battery cleaning tool. Use the battery post brush end of the cleaning tool to polish posts.

■ A terminal spreader to facilitate putting terminals back on posts. Do not bang the terminals into place with a hammer. You may ruin the battery.

With terminals on battery posts, tighten them securely and spread a light coat of petroleum jelly over them to help retard corrosion.

**wrong torquing**

3. The biggest problem that has been encountered with side battery terminals has been improper torquing, which causes high resistance and hard starting. To service side terminals properly, do the following:

■ Unscrew the negative battery cable.

■ Clean the threads and end of the cable screw thoroughly with a wire brush.

■ If necessary, clean the internal threads in the battery case. Use a small wire brush or a ⅜-16 bottoming tap. If you use a bottoming tap, do not tighten it beyond 70-inch-pounds.

■ Service the positive cable next.

■ Install the positive cable and tighten it to 70-inch-pounds. Then install the negative cable.

4. You should replace a battery cable when insulation wears or cracks, or wires become exposed around terminals. Note the size of the cable you are buying to make sure it's the same size as the old cable.

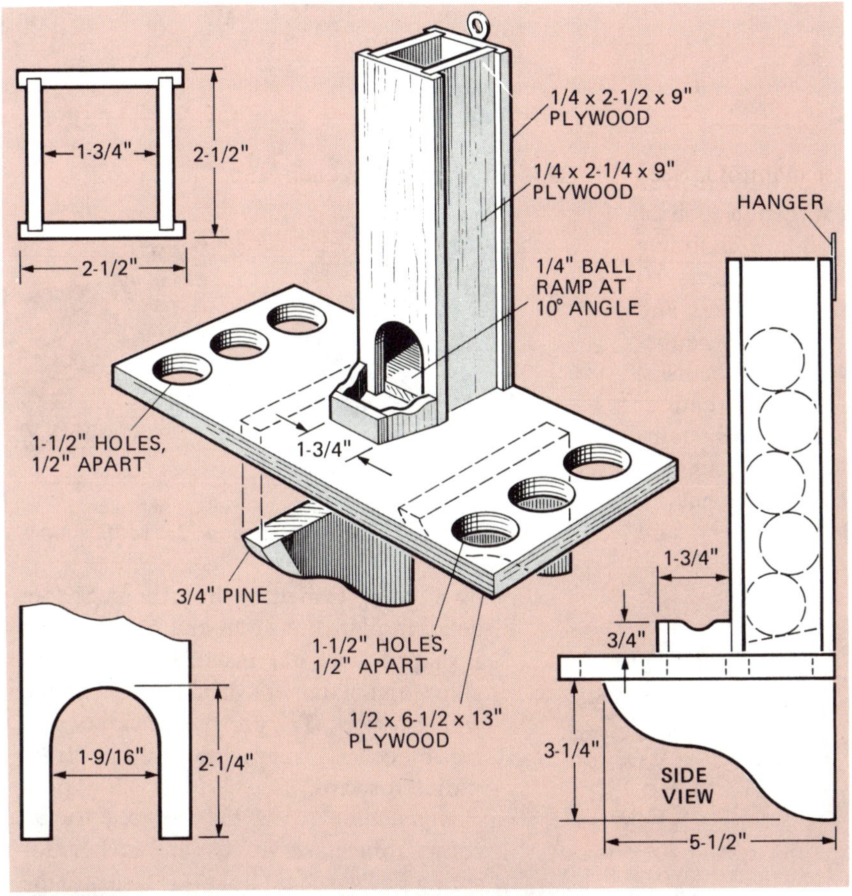


# Ball and paddle caddy

By NORMAN WARD

■ THIS NEAT wall rack holds six table tennis paddles and stores seven balls. The balls are stored in a chimney-like dispenser that lets them roll out one at a time from an opening at the bottom.

You can make the rack of plywood leftovers from your scrap box or, for a fancier appearance, from walnut. The holes for the paddles are bored with a 1½-in. spade bit, and you'll do the neatest job if you clamp a scrap board to the underside of the plywood. A slanting block in the bottom of the ball holder makes the balls roll forward.

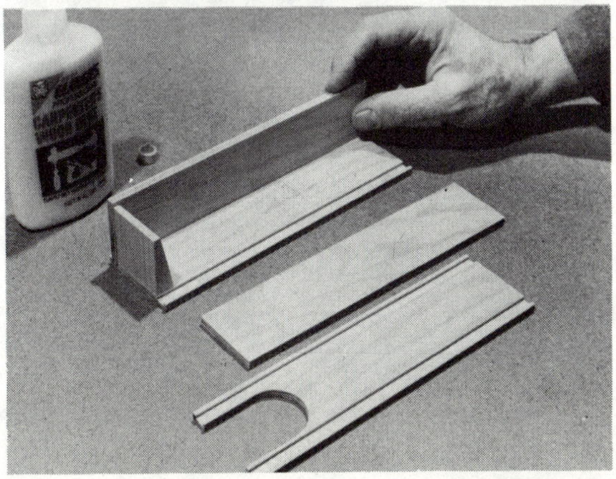

**SIDES OF BALL HOPPER** fit in grooves cut in front and back. Run grooves first, then cut ball opening in front. Insert slanting block in bottom as you glue the parts.

SEE ALSO

**Air games . . . Backgammon tables . . . Bridge sets . . . Bridge tables . . . Chess sets . . . Cribbage boards . . . Family rooms . . . Game tables . . . Pool tables . . . Shuffleboard tables**

# Tool caddy from scraps

■ REPAIR JOBS around the home often require trips back to the workshop for forgotten tools or supplies. The tool caddy here should eliminate such wasted time. It not only totes a wide assortment of tools, but also organizes fasteners and spare parts. Since the organizers are interchangeable, you can fit out the caddy for specific tasks.

Construction of the caddy is simple, but attention must be paid to the clever latches and retainers for the saws and hammer. The hammer, which also serves as the caddy handle, is held in place by pipe strapping at the handle end and by a spring-loaded latch at the neck (see detail in plans). A strap and buckle or an L-shaped steel bar installed in slots cut above the hammer's neck in the cradle will also work. The protruding end of corner brace (U) keeps the hammer from sliding forward.

A machine screw (M), brazed to the corner brace, serves as an anchor for spring F. Two chain links (G) fasten to the spring and engage the blade-tensioning bolt on a hacksaw. A similarly devised hook secures the backsaw at the other end of the caddy. Instead of chain links, a hook bent from hanger wire engages the saw handle.

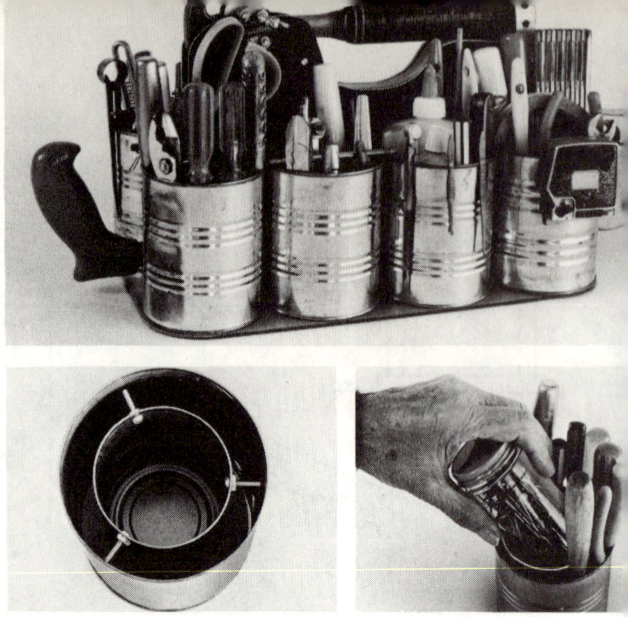

**INGENIOUS TOOL CADDY** organizes hammer, saws, rulers, tape, and other tools as well as hardware. Three bolts, 120° apart are spacers. Jar holds supplies.

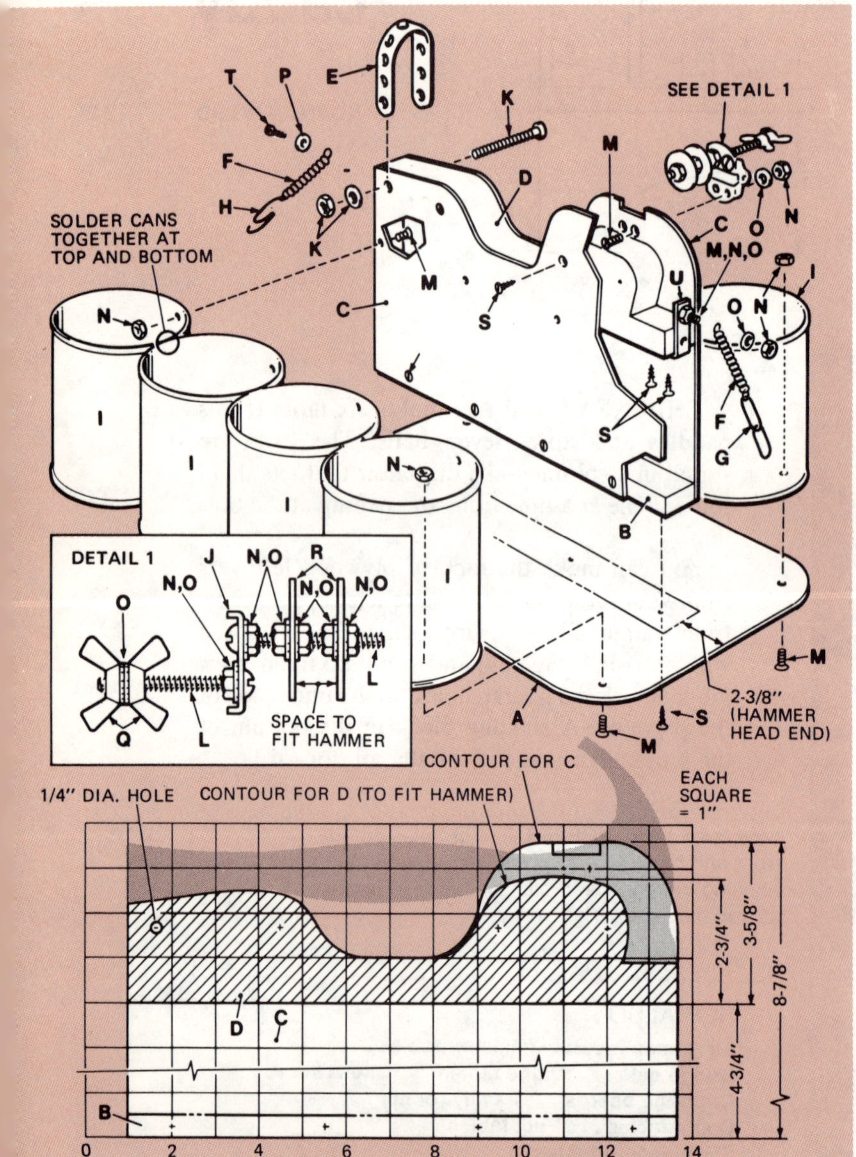


PATTERNS FOR PARTS C AND D

**MATERIALS LIST—TOOL CADDY**

| Key | No. | Size and description (use) |
|---|---|---|
| A | 1 | ⅛ x 9⅜ x 16¼" hardboard (caddy base) |
| B | 1 | ½ x 1 x 12⅝" pine (saw-holder base) |
| C | 2 | ⅛ x 8⅞ x 12⅝" hardboard (saw-holder sides) |
| D | 1 | 1 x 2⅜ x 12⅝" overall (hammer cradle); sandwich ¾" pine between 2 pieces of ⅛" hardboard |
| E | 1 | 3" pipe strapping (handle retainer) |
| F | 2 | ⅛"-dia. x 2" spring |
| G | 1 | 2-link piece of chain (to catch on hacksaw) |
| H | 1 | hook bent from stiff wire (to catch on backsaw) |
| I | 8 | 4" dia. cans |
| J | 1 | self-closing cabinet hinge |
| K | 1 | ¼-20 x 1½" machine screw, nut and washer |
| L | 2 | No. 8-32 x 1¼" machine screw |
| M | 15 | No. 8-32 x ½" machine screw |
| N | 22 | No. 8-32 hex nut |
| O | 11 | No. 8 external-tooth lock washer |
| P | 1 | flat washer |
| Q | 2 | No. 8-32 wingnut |
| R | 2 | 1" o.d. flat washer |
| S | 20 | ⅝" No. 8 fh screw |
| T | 1 | No. 8 pan-head screw |
| U | 1 | corner brace |

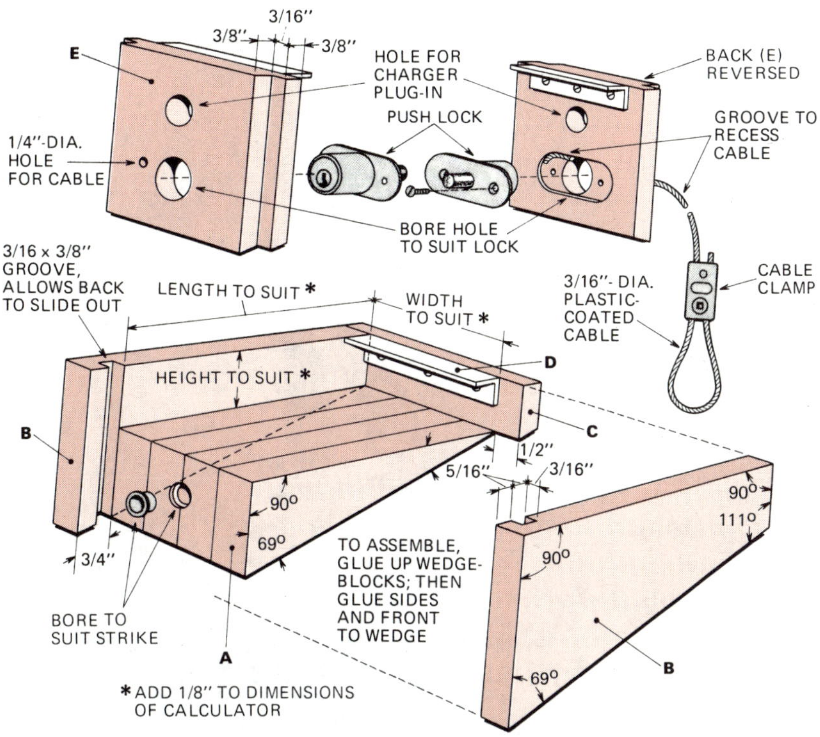


# A stand for your calculator

**WHEN STAND** is unlocked, back of calculator can be raised for removal. Suit dimensions to your calculator.

■ POCKET CALCULATORS have a tendency to "grow legs and walk off" without a stand such as this one.

Begin by tracing the side (profile) of your calculator. Add ⅛ in. to the overall length and height. Next, draw in the area for the bottom wedges and back and front panels. Use this diagram as a pattern to cut out sides B and wedges A.

Following the drawing, cut the wedge and side sections, glue together, then glue on the front. Bore a hole for the lock in the back. Then mortise the lock flush. Drill a hole through back of lock faceplate for cable and cut a small groove under the faceplate to hold it.

Bore holes in the back large enough for the lock and charger plug. Cut and drill two pieces of aluminum angle. Each should be attached flush with the top of the case.

**STAND MAKES** calculator easier to read. Cable prevents theft.

**MATERIALS LIST—STAND**

| Key | No. | Size and description (use) |
|---|---|---|
| A | 4 or 5 | ¾" pine; short side of wedge is 1¾" long; other dimensions to suit (base) |
| B | 2 | ½" hardwood to suit (side) |
| C | 1 | ½" hardwood to suit (front) |
| D | 2 | ½ × ½" aluminum angle to suit (retainers) |
| E | 1 | ¾" hardwood to suit |

**Misc.:** Push lock*, cable clamp, 3/16"-dia. plastic-coated cable × length to suit

*Chrome push lock No. 524 (for sliding doors and panels) from Lustre Line Products, Philadelphia, Pa. 19125, was used in prototype. It costs about $5.50 at hardware stores, building supply outlets and retail locksmith shops.

SEE ALSO

**Bookends . . .Bookracks . . . Clocks . . . Desk accessories . . . Desks . . . Drawers . . . Finishes, wood . . . Gifts, Christmas . . . Joinery . . . Letter racks . . . Weekend projects**

# Make your own camera harness

By KEN GALLARD

SKIER WEARS no vest or jacket to show harness in action; normally he would wear one to keep the cameras warm and protected.

A CUSTOM-MADE harness takes the worry out of bringing camera equipment along for activities such as skiing or motorcycling. This one is made from 8-ounce pack cloth, nylon webbing and elastic straps.

Sewn between two layers of cloth are a piece of ¼-inch, closed-cell foam for each camera and two 2-inch webbing strips running top to bottom. Ultraheavy-duty, 1-inch elastic straps are sewn to the outside to hold the cameras in place. Two-inch, 9000-pound-test webbing is used for the back straps; buckles and 1-inch webbing for the adjustable side straps. Another 1-inch strip may be used as a stiffener along the collar front.

A standard fanny pack can be modified to carry lenses and other photo gear. With a hand-stitching awl, sew in a layer of ½-inch, closed-cell foam all around. Lenses are held in place by Velcro-closed nylon or elastic straps. The opening straps have large loops for handling ease in the cold.

All materials, including awl, are available from camping or mountaineering supply companies. If you don't need all this equipment, the design is easy to adapt to your needs.

SEE ALSO
**Camera bags . . . Camera care . . . Camera gunstocks . . . Cameras . . . Photography . . . Used cameras**

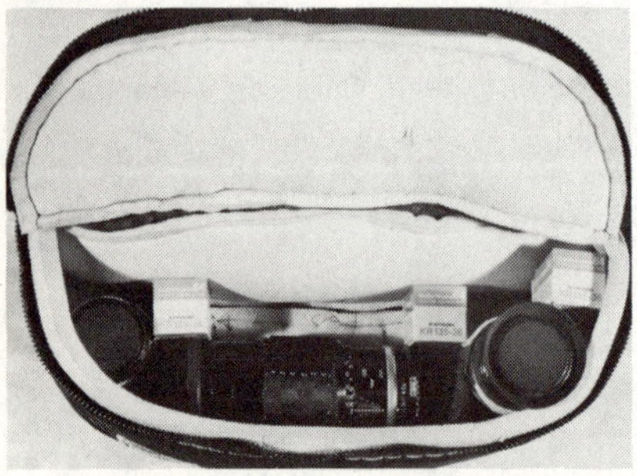

ALL SIDES of any pack are lined with foam. Straps close with Velcro.

CAMERAS rest on ¼-inch foam sewn between cloth layers.

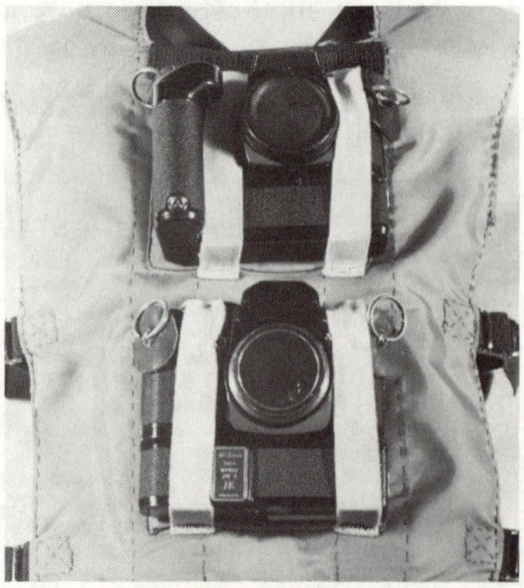

9000-POUND-TEST webbing provides support and distributes the load.

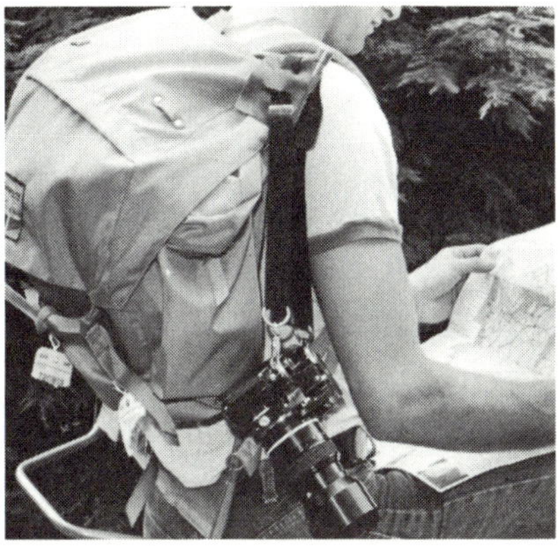

**HANG** your camera from the frame of your backpack when hiking. This distributes camera's weight better and avoids sore neck muscles.

**A 2½-GALLON** plastic bucket is useful for mixing large amounts of photographic chemicals. Its size allows vigorous stirring without spills, and its pour spout makes bottle filling easier.

**WITH AGE,** a camera bellows will develop holes that leak light. Brush-on plastic tape quickly repairs such leaks; it's lightproof, flexible, easy to apply and quick to dry.

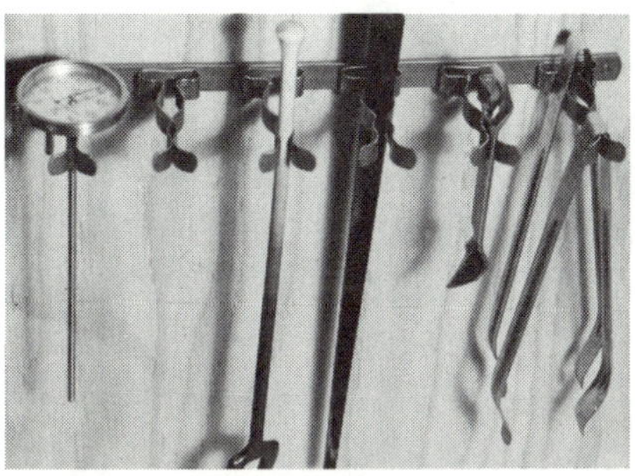

**A TOOLHOLDER** organizes miscellaneous darkroom items. It stores them safely, but keeps them handy.

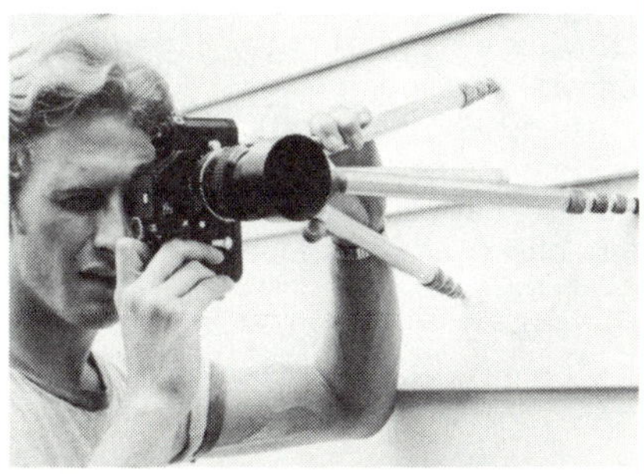

**SLOW SHUTTER** speeds require use of a tripod, but often a tripod takes too long to set up or isn't tall enough. Try bracing it unextended against a wall; the camera will be supported as rigidly as if the tripod were on the floor.

**INSTEAD** of hanging a hand-held light meter from your neck, fasten the meter to the chain of a Key Keeper and clip the keeper to your belt.

**CAMERA REPAIR** is a tricky business under the best conditions, but if a small part gets lost, you may as well give up. Avoid the problem by keeping parts on a glass- or plastic-covered dish, like those used to serve cheese.

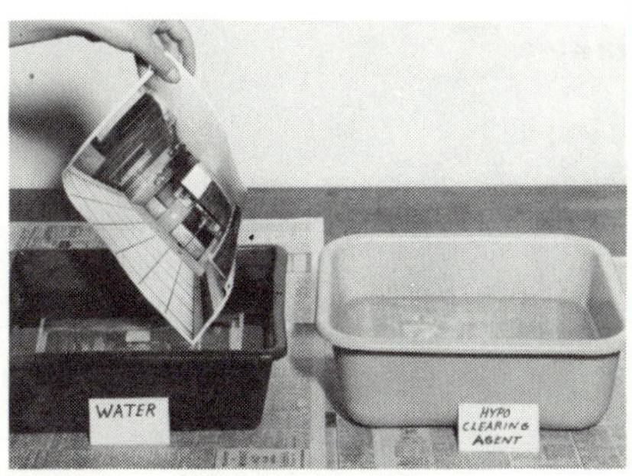


**PLASTIC DISHPANS** make inexpensive deep trays for print-processing solutions. They are large enough for 8 x 10s and will hold many prints at once.

**THE METAL** hardware used to join camera strap and camera can chip the paint on the camera body. A leather disc with a hole in it slipped over the strap mount cures the problem.

# Two ways to make a custom camera bag

## LINE AN ATTACHE CASE WITH CARDBOARD

By GLOVER HENDRICKSON

**TIGHT, EFFICIENT PACKING** is easy with cardboard separators trimmed into compartments to fit equipment. And it's sturdy. For three years, this bag has carried two 35-mm cameras, three auxiliary lenses, 10 rolls of film, flash equipment, 30 flashbulbs, rangefinder, light meter, filters, tape recorder, nine cassettes, five sets of spare batteries and extra power pack for the recorder.

■ WHAT DOES IT TAKE to hold all this gear neatly and firmly in a slim attaché case?

Just corrugated cardboard. Cut a plain, flat sheet of it to form a box two inches deep, just fitting the case. Arrange the equipment as compactly as possible and mark its outlines. Then cut within the outlines with a sharp knife, leaving some excess as a flap that can be turned under

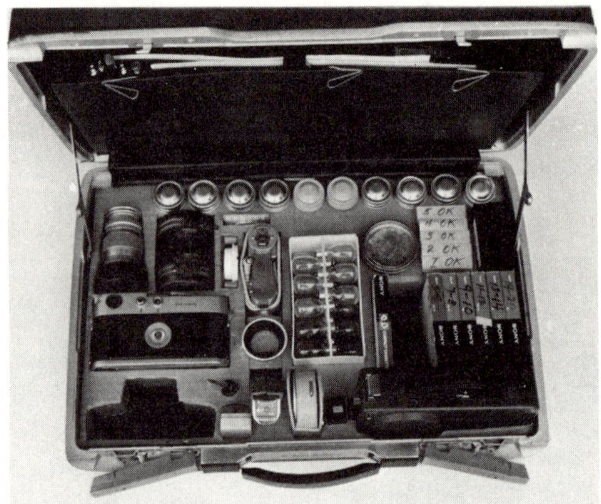

## LINE YOUR SUITCASE WITH SOFT FOAM

By MICHAEL LAMM

SEE ALSO

**Camera accessories . . . Camera care . . . Cartridges, film . . . Darkrooms . . . Outdoor photography . . . Photography . . . Slides, photo . . . Used cameras**

**FOAM PADDING** is great protection against jars and jolts of travel. Large scissors cut it easily, whether you're cutting all the way through for a big camera or just making a shallow indentation to hold light meter or strobe. Old suitcases are best—they're less inviting to thieves.

■ OUTGROWING YOUR GADGET bag? Here is a cheap, satisfactory alternative: Just convert an old suitcase to a foam-padded carry-all.

First scrounge your suitcase; most families have a few kicking around the attic. It needn't be fancy, but should be sturdy, and big enough to hold all your equipment plus a fat, foam surround.

The foam comes from surplus stores and similar outlets; the dealer can cut it to the interior dimensions of your suitcase. You'll need one-inch slabs for the top and bottom of the case, plus another slab thick enough to fill the space between them (i.e., if your case is six inches deep, you'll need a four-inch center section to fit between top and bottom pads).

Arrange your equipment on this center section and trace its outlines with a felt-tipped pen. Then cut holes and indentations with a large sewing scissors. If your suitcase has pockets, you may want to leave them exposed to hold film. The finished case will protect your camera and accessories from drops and jolts of all kinds.

and glued to the partition reinforcements.

Turn the box over and partition the individual compartments with two-inch strips of the cardboard, then roll or work more cardboard into the voids between the partitions (the picture on the right shows the partitions before the voids were filled). For additional protection, you may want to line your attaché case, top and bottom, with foam.

If you want to carry more gear, partition both the top and bottom of a valise of suitable size with foam-covered cardboard, separating the two layers so your equipment won't bang together and become damaged.

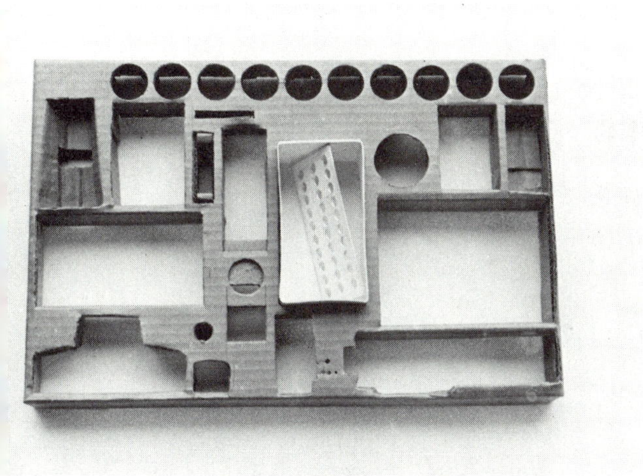

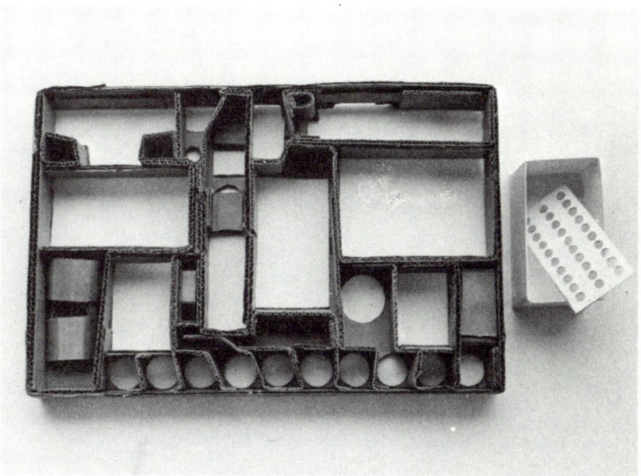

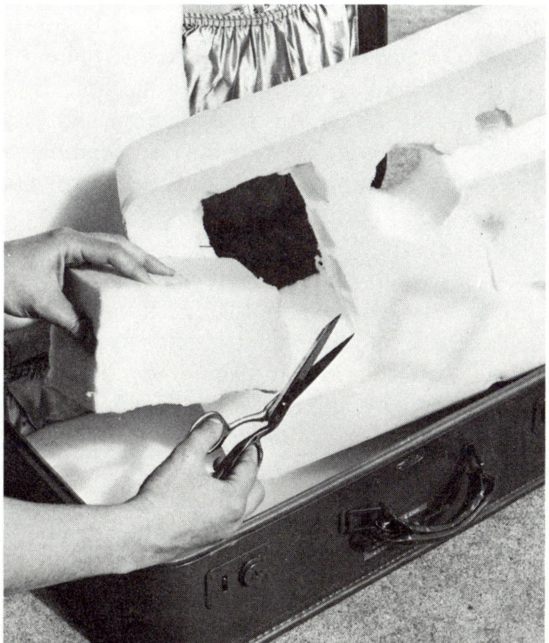

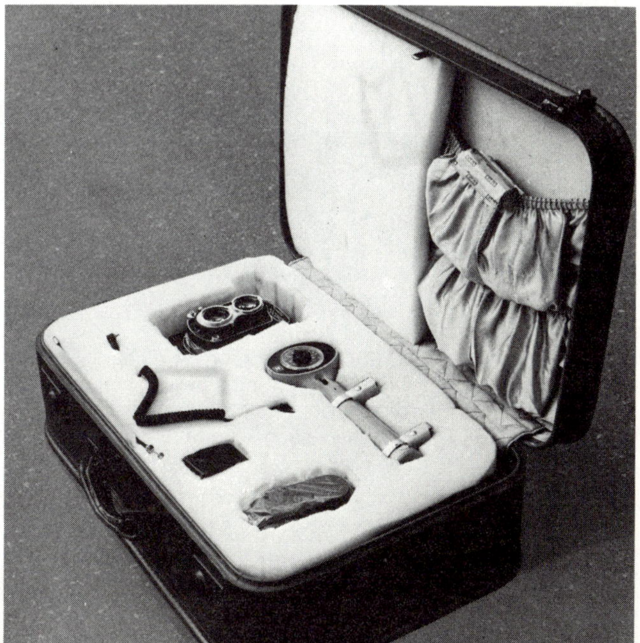

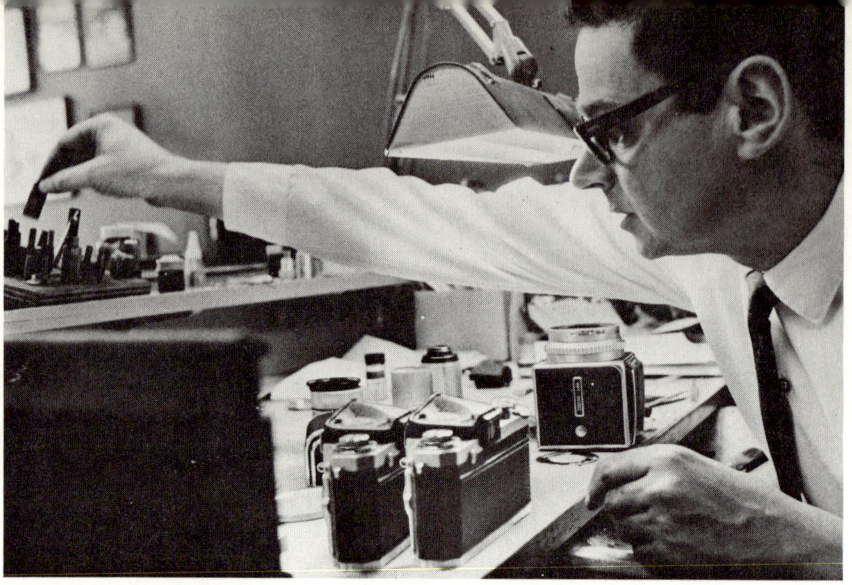

By PHIL GERACI

**EXPERT REPAIRMAN** Charlie Scheer, shown in his Washington, D. C., shop, keeps cameras rolling for the nation's top press photographers who cover the White House. His suggestions for good camera care are shown on these pages.

# Simple checks to keep your camera shooting

**The smart way to care for your camera, says a leading repairman, is to know how to wage effective warfare against its No. 1 enemy—dirt.**

**GLASS WINDOWS** for viewfinders, range finders and exposure meters often become clouded with grime. Keep them clean with a cotton swab and alcohol to dissolve film deposits.

SEE ALSO

**Camera bags . . . Cameras . . . Darkrooms . . . Developing, photo . . . Lightboxes . . . Photography . . . Prints, photo . . . Slides, photo . . . Used cameras**

EIGHTY PERCENT of all camera troubles are caused by dirt. That's the word from one who should know. He's Charlie Scheer, owner of National Camera Repairs in Washington, D. C., and chief consultant on camera problems to top White House news photographers.

Since dirt is a camera's No. 1 enemy, it follows that getting rid of it is the No. 1 step in good camera care. This is especially important when photo equipment has lain around unused for a time, as is often the case during the winter months. Fine dust particles can sift into a camera just sitting untouched on a shelf. Oily fingerprints and moisture attract and hold dirt like glue so that grimy deposits quickly build up.

If you just haul out your camera and start banging away with it, you grind the accumulated dirt deeper into the controls and delicate mechanisms. Your shutter starts to slow down, your flash synchronization goes out, the film advance may jam, and suddenly you're in trouble.

The most likely spots for dirt to seep into a camera are around the controls. These should be kept clean by regular brushing with a soft camel's-hair brush. Use a small, good-quality paintbrush or a photographer's lens brush. Built-up grime that can't be brushed away can be removed with a cotton swab dipped in rubbing alcohol. Also use a cotton swab and alcohol on the windows for viewfinders, range finders and built-in exposure meters.

Delicate parts like the lens, the film pressure plate and the mirror on reflex-type cameras re-

**KEEP LENSES CLEAN** with a soft brush. Use a special photographer's lens brush for this. The one shown above combines a bristle tip with a rubber syringe for blowing a stream of air, useful after brushing.

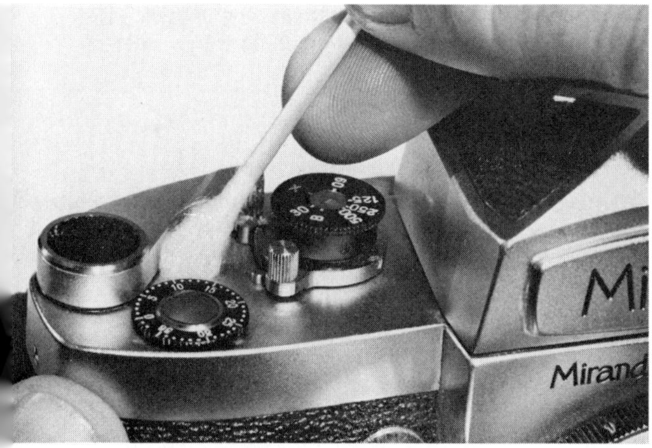

**DIRT AND GRIME** collect around a camera's controls and can eventually work their way inside, gumming up intricate mechanisms. Remove such deposits with a cotton swab dipped in rubbing alcohol.

**A VACUUM CLEANER** is handy for removing dust from hard-to-reach places in the camera. Use it only on suction, though; never when it's on blow. A blast of air can force dust particles deeper into critical parts or add new dirt, complicating the cleaning task.

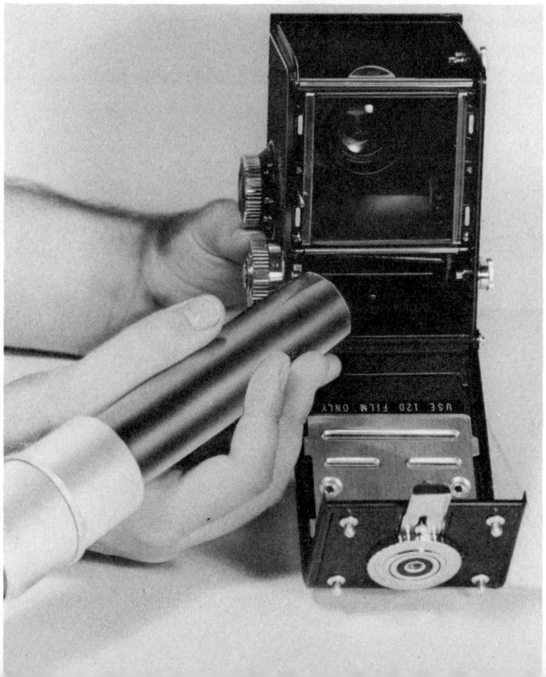

**A DELICATE MIRROR** used in a reflex-type camera requires special handling. You can brush it lightly if necessary, but it's even better to use only a gentle air stream from a syringe, as shown below.

**BLOW DUST** from inside the camera with a small rubber syringe. If the mirror is the type you can lock up, clean it first in the down position, then raise it out of the way and blow out the dirt behind it.

quire special care. The pressure plate is found on the inside of the camera back and its function is to keep the film flat and smooth behind the lens. You can brush gently around it, but be very careful not to scratch it. A scratch on the plate will, in turn, scratch the film pressed against it.

When working with lenses and mirrors, Scheer advises extreme caution. Use the least amount of cleaning that will do the job. Scheer believes that most people overclean their lenses while ignoring the rest of the camera. Excessive wiping can damage the thin antihalation coating on a lens and cause scratches. Start by blowing dust particles away with a small rubber syringe. These are available at camera shops and drugstores. Dirt that can't be blown away can be loosened by gentle brushing. On removable lenses, clean the rear element as well as the front one.

If necessary, the blowing and brushing can be followed by a light wiping with lens-cleaning tissue to remove any remaining film. Use only the special tissue made for this purpose, however—never a handkerchief or other cloth as the threads can cause scratches.

On mirrors, use only a combination of blowing and gentle brushing. Never touch or wipe a mirror or you may damage its precise alignment. Also never remove a lens and leave the opening in the camera exposed. Cover the hole with a body cap or keep a lens on the camera at all times.

Dirt inside a camera is even worse than on a lens. Open the back and shake the camera first to dislodge any bits of film or paper that may have been torn off around the sprockets. Then brush out the interior, being very careful around the pressure plate and the thin curtain on cameras with focal-plane shutters. Follow this with blowing to remove particles loosened by the brush. Always use a syringe for this. Never blow with your mouth or moisture in your breath will collect in the camera and cause rusting.

A vacuum cleaner is handy for removing dust, but use it only on suction. If you blow with it, you'll blast more dirt into the camera than you take out. Be careful not to touch anything inside the camera with the metal nozzle. Never attempt to oil any of the mechanisms, either, warns Scheer. This is a job for a professional.

After cleaning, the next thing to check on is your shutter. Shutter speeds generally slow down as dirt deposits build up. Run a test roll of film through the camera at different exposures. Try some shots with and without flash. The exposures that are indicated as correct for your test scenes should produce good results, and there should be an even gradation of light and dark pictures on either side of the correct exposure.

Also try a series of tests at the same exposure but using different combinations of shutter speeds and lens openings. Open up the lens a stop at a time as you use progressively faster speeds. If your shutter is working properly, all of your test results will be exactly the same since the exposures are identical. If they aren't, it's an indication that your timing is off.

If there are signs of trouble, take your camera to a professional repairman and ask to have the shutter adjusted. Don't attempt to do this job yourself.

**TINY SCREWS** in lens barrels and other camera parts frequently work loose and fall out, causing serious damage. Give them a periodic tightening with a small jeweler's or model-maker's screwdriver.

**USE LIQUID LENS** cleaner only as a last resort to remove stubborn gummy deposits. Swab it on gently with a wadded-up lens-cleaning tissue, then wipe the lens dry. Excessive use of liquid cleaner can harm a lens's coating.

**A SOFT-BRISTLE** paintbrush of the narrow sash and trim type is useful for getting into crevices inside a camera. Brush out the film compartment, but don't damage the fabric curtain on cameras with focal-plane shutters.

**TRY THIS QUICK** test for proper flash synchronization on cameras with focal-plane shutters. Point the camera at a white wall, open the back and trip the flash. The sync is good if the opening through the lens appears round (top photo). If the opening is partly blocked by the shutter curtain (lower photo), the sync is off.

**ADDING A SKY** filter to your lens will protect it from fingerprints and scratches. Such ultraviolet filters have no effect except to reduce the haze on distant shots, require no adjustment in exposure and can be left on a lens permanently.

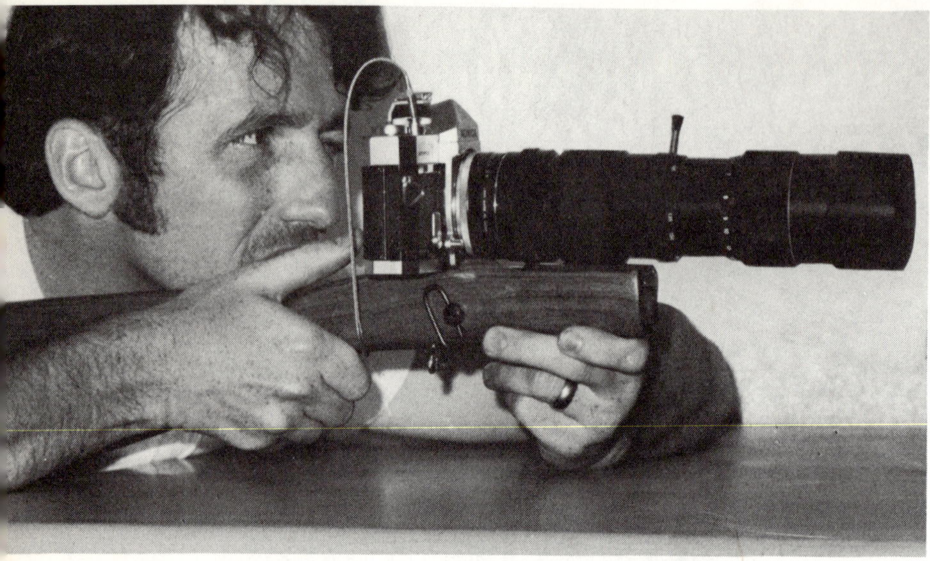

**STEADY, AIM, SHOOT!** A gunstock keeps your telephoto shots sharp. And a stock from a real gun (this one's from a Mauser) is easy to adapt for camera use. Note how forearm has been cut to clear zoom lever on the lens.

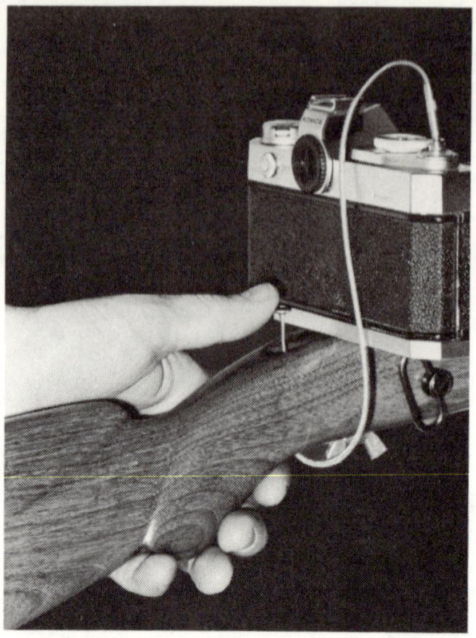

**CABLE** release makes it easier to hold stock steady as you squeeze off shot. Mount as shown, or from below.

# Inexpensive telephoto gunstock

**By BOB GREWELL**

ARE YOUR telephoto shots blurry? A gunstock will steady your camera without tying it down as a tripod would. And though commercial camera stocks run about $30 and up, you can build one easily and inexpensively.

Start with a used or unfinished gunstock (about $5 and up at gunsmiths). Check its size by taping your camera in place and holding it. You can cut down the stock (after allowing for thickness of the butt plate) or add a thicker butt plate as needed.

The camera mounts with a bolt through the stock's trigger slot. Cut a ¼-inch metal plate a bit longer and wider than the slot and drill a 5/16-inch hole in it for the camera bolt (you can try a large, heavy washer if no means are handy for cutting metal).

If you can find a ¼-20 thumbscrew long enough to reach the camera through the slot, use it. Otherwise, force-fit, braze or epoxy a wingnut to the head of a ¼-20 bolt. Start with a bolt a trifle long, then grind it back, a bit at a time, until it just holds the camera snugly by its tripod socket.

SEE ALSO

**Bird photography . . . Boat photography . . . Camera bags . . . Candid photography . . . Montages, photo . . . Night photography . . . Outdoor photography . . . Panorama photos**

Drill a hole down through the stock for a cable release. You can run the release up from under the stock, but I find it more convenient to press it from the top with my thumb. Now all that's left to do is to finish or refinish the stock. You can also add sling swivels, if you like.

**SCREW AND** mounting plate, inserted from under the stock, hold the camera by its tripod hole.

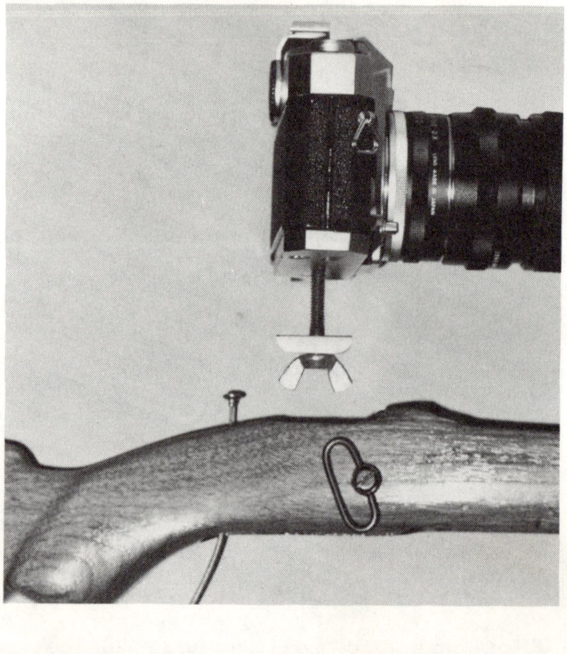

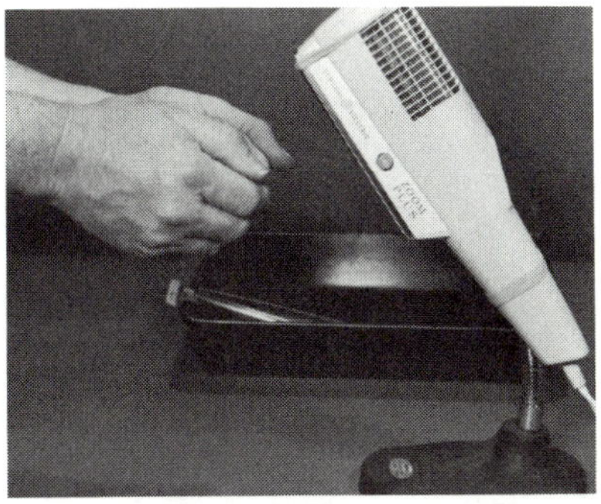

**CLEAN, DRY HANDS** are necessary to make sure enlarging paper doesn't become contaminated in the darkroom. A hair dryer fastened to a gooseneck or other stand allows you to hold your hands in the airstream for drying.

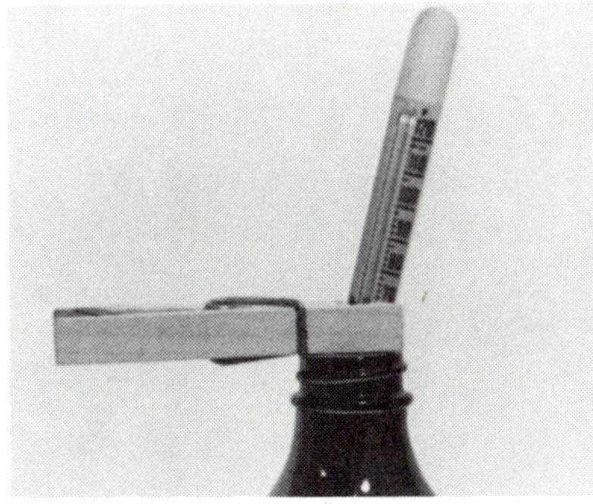

**A CLOTHESPIN** will keep a thermometer from falling into a tall bottle when you take temperature readings.

**TINY SCRATCHES** on a negative are big lines on an enlargement. A tiny amount of petroleum jelly applied to the negative with your finger will fill the scratch and keep it from showing on the enlargement.

**A SMALL CAMERA** may be supported on many surfaces with a spring photo clamp. The swivel-bolt device is fitted with a bolt, which is then screwed into the tripod socket.

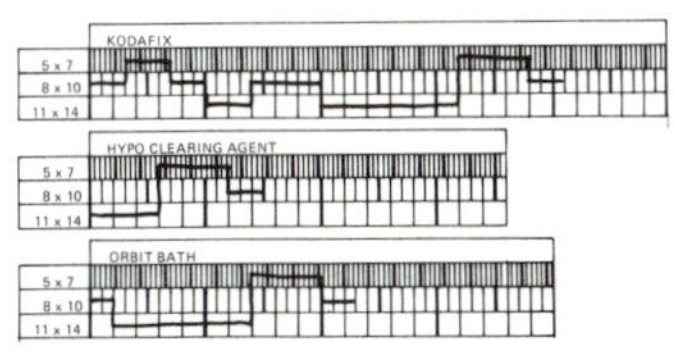


**THIS CHART** makes it easy to plot the depletion of photo chemicals. Using manufacturer's specs, compute the number of prints of each size obtainable from each batch of chemicals. Draw chart showing each print capacity and mark off the number of prints of each size as they are sent through each bath. When chart indicates solution is exhausted, replace with fresh batch.

**WHEN USING** your camera's self-timer or when making long exposures, light may enter the camera through the eyepiece if you're not looking through the viewfinder. To eliminate the problem, cover the eyepiece with black tape after composing and focusing.

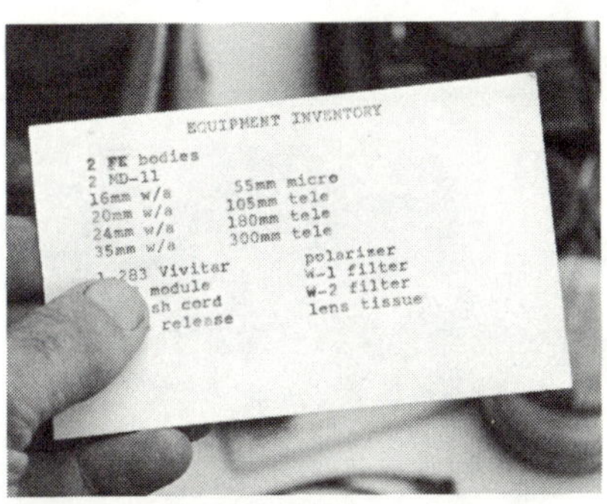


**WHEN PACKING** or repacking photo gear, it's easy to leave behind a minor accessory or even a lens if you're in a hurry. An inventory card kept in your bag lets you check off each item as it goes in.

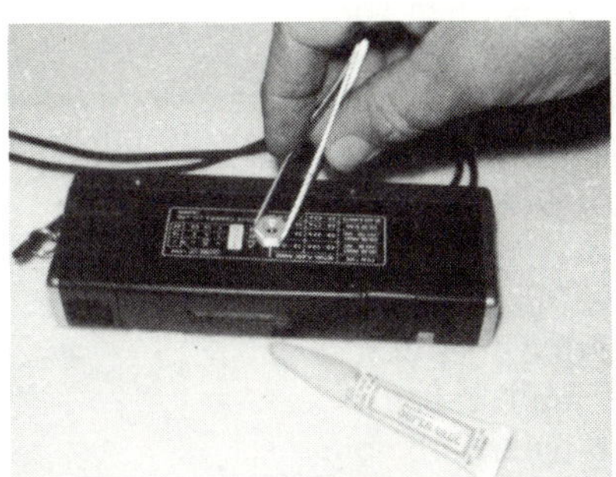

**A ¼-20 NYLON NUT** glued to the base of a 110 camera makes it suitable for use with a tripod. Cyanoacrylate glue works well.

**TAPE** a ¼-inch rod, slightly off center, to the underside of a developing tray. When prints are in the solution, set the tray rocking. The liquid will surge to and fro several times before stopping. By repeating this agitation every 10 seconds or so, prints will be more evenly developed.

RANGEFINDER CAMERA

# What type of camera should you buy?

**Instamatic, rangefinder, twin-lens or single-lens reflex? What camera is the best for you? A better camera can make you a better photographer—but only if it helps you get shots your present camera can't take. Here are some tips to tell you when you need a better camera**

TWIN-LENS REFLEX

■ THE TIME to buy a new camera is when your present camera cramps your style—and not before. Within limits, the cheapest Instamatic, used with care, can shoot as good a picture as the most expensive Nikon, Hasselblad or Leica used with equal care.

But there are a lot of pictures that an Instamatic just can't shoot at all. And that doesn't just apply to simple cameras: The most expensive camera in the world will sooner or later stub your toe on its limitations (though often there's an ac-

SEE ALSO

**Bounce-flash photography . . . Camera care . . . Cartridges, film . . . Dryers, photo . . . Filters, photo . . . Montages, photo . . . Photography . . . Prints, photo . . . Slides, photo . . . Used cameras**

SINGLE-LENS REFLEX

**AVAILABLE-LIGHT** capability to shoot indoors without flash requires fast lens and fast films.

**TIME EXPOSURES** of several seconds can yield night shots like this. Night shots can even look like daylight.

**FAST ACTION** requires fast shutter speed to prevent blur. Speeds of 1/250 or 1/500 usually suffice.

35-mm negative size

**SIZE MAKES** a difference: 35-mm negative requires more enlargement (for the same size image) than 2¼-in.-square, 120 image, and the difference clearly shows.

2-¼ x 2-¼″ (120) negative size

**SHARPER,** crisper print from 120 negative is closer to contact-print size, hence is also closer to contact-print quality.

**WIDE-ANGLE** lens emphasizes perspective and has great depth of field, lets you get wider area.

**TELEPHOTO** lets you shoot "close-ups" when you can't get close; compresses space in the distance.

**MACRO** lenses and a wide range of other close-up accessories let you make small subjects big.

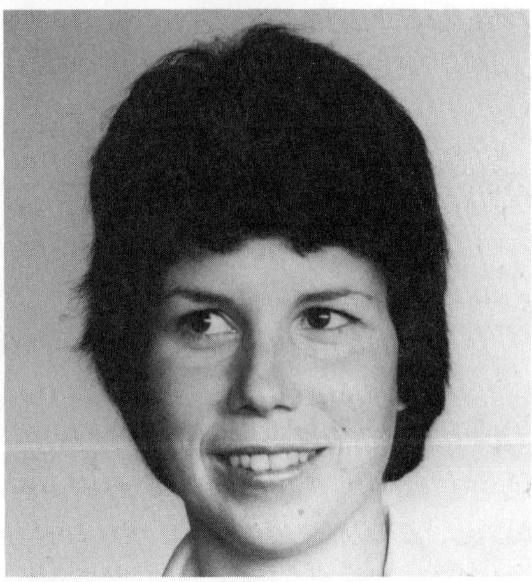

**PORTRAIT WITH** normal lens (top photo) tends to make the subject's nose balloon toward the camera, unless you move away till the subject no longer fills frame. The portrait with the mild telephoto (bottom photo) has the subject filling frame, even though shot from far enough away for pleasing perspective.

cessory available to un-stub your toe, again). A fine camera gives you all kinds of latitude and flexibility that you can't get with a simple camera. Expensive cameras are complete photo *systems.*

Before you run off to buy a camera that will do more for you than your present one, make sure you need to do more. Determine what you want in a photograph and get the camera that will do the job. Here's a guide to what better cameras and their features can do for you, and to tell if you need them.

■ **Dim light is a problem** that most under-$100 cameras cope with poorly, but which most over-$150 cameras handle well. Under-$100 cameras have "slow" lenses with apertures of about f/8 to f/11, which don't let much light in to reach the film. Better cameras have lenses of f/2.8, f/1.8 or faster, which let in at least eight times as much light, letting you shoot when it's eight times as dark.

**slow shutter speed**

Less expensive cameras may also have other dim-light disadvantages: Their shutters may operate no more slowly than 1/30 or 1/45 second (the more time the light has to reach the film, the more light will reach it). And if they only take 110 or 126 cartridge films, they can't use such fast films as Tri-X, which don't need as much light. Thus you can't freeze any really fast action with an inexpensive camera because there's no fast film to fit it.

■ **Action foils** the cheaper cameras in other ways, too. Stopping action takes fast shutter speeds, which many of the least expensive cameras lack. Simplified electric-eye cameras may have fast speeds available—but their electric eyes are probably programmed to use those fast speeds only when the light is very bright. More sophisticated cameras let you pick either a shutter speed to match the action or a lens aperture that will force your camera to give you the shutter speed you want.

The less time the light has to reach the film, the more rapidly you should let it in. So a wide aperture (at least f/2.8) is advisable for any camera you intend to use for action photography.

■ **Selective focus** is another trick that better cameras will let you try. A simple camera's f/8 or f/11 lens keeps everything in focus from about five feet to the far horizon. That's fine when you want both near and far objects in focus; but if you want to throw a background out of focus while keeping the nearby subject sharp, you'll need a camera with a fast lens again. At apertures of f/2.8 or faster, only a narrow zone is in sharp focus.

Since focusing gets more critical with faster lenses, better cameras have some means of checking focus accurately. On cameras list-priced from about $125 or $150 up, this is usually a rangefinder; on reflex cameras, it's a ground-glass or similar focusing screen.

■ **Rangefinder cameras** show your subject through a window with a fuzzy image in its center. That fuzzy spot is actually two images, slightly out of register; when the two images overlap exactly, the fuzziness will disappear and the lens is in focus.

Rangefinder focusing is easy after a few minutes spent getting the hang of it, and is quick. Rangefinder cameras, in addition, tend to be compact, quiet and inexpensive.

■ **Reflex focusing** is closer to what actually takes place on the film: The image will be sharp when and where it looks sharp in the finder, fuzzy when and where it doesn't. Some people find this easier to use and understand than a rangefinder. Others find it harder because it forces them to judge small differences between the sharp and not-quite-sharp. For them, most single-lens reflexes have either a "microprism" spot that slightly exaggerates the difference between clear and slightly fuzzy, a "rangefinder" spot that breaks up out-of-focus lines or both. Twin-lens reflexes, with bigger ground-glass images, tend not to have these extras.

■ **Visualizing your image** is important. If you can't relate what you see in the finder to what you'll see on your print or slide, you can't expect good pictures. For many people, reflex finders look more like a picture, less like a peephole on the world, and make this visualization easier.

**bigger screen and twin-lens reflex**

This effect is especially true of the big ground-glass screen on a twin-lens reflex. That screen is big because the film is big, and the bigger the negative, the higher the potential print quality. Twin-lens cameras are easily used at waist level or below, which makes them great for child photography from the child's eye-level and viewpoint.

The single-lens reflex, or SLR for short, is far more popular, because it can do just about everything, and do some things uniquely well. The key to its versatility is its viewing system, which lets you look through the camera lens to see just what the film is seeing (when you shoot, the viewfinder mirror flips up to let light through to the film).

That makes the SLR exceptionally good for close-up work, or for use with interchangeable lenses.

No matter how close you get with an SLR, you don't have to worry about cutting off part of your subject because your viewfinder is in one place and your lens in another, as is the case with a twin-lens reflex. And you can see whether focus is correct or not.

**see what film sees**

For the same reason, no matter how long a telephoto lens or how wide a wide-angle you use, you will see what the lens and film see, and be able to focus without cumbersome auxiliary equipment.

Being able to take off one lens and mount another makes a vast difference in what a camera can do for you. Wide-angle lenses, for example, don't just let you get more in the picture, important as that is when shooting in cramped places; they also let you exaggerate perspective and dramatize space.

Long telephotos, on the other hand, compress space, making cars spread over a mile of freeway look bumper-to-bumper; and long teles give you "close-ups" of subjects too far away to reach (or altogether inaccessible, like the helicopter on page 523).

Medium teles, from 85 to 105-mm, let you shoot portraits from the distance that gives the most flattering perspective, yet still fill the frame with your subject.

An SLR camera equipped with suitable wide-angle and telephoto lenses gives you the biggest range of options. But do you really need all those options?

**SLR's advantages, disadvantages**

So should you buy an SLR? Yes—if you want to use its versatility and are willing to pay the SLR's high cost and the cost of additional lenses or close-up accessories. You might even want one just because you like its viewfinder system better than that of other camera types. On the other hand, realize before deciding that the SLR is noisier, more complex, more expensive, and usually heavier than other types, and the viewfinder blacks out for an instant every time you shoot (it's that flipping mirror). The biggest of these disadvantages are cost and weight. If you only want to take snapshots on vacation trips, you're probably better off with an inexpensive pocket camera.

■ **Buying a new camera** won't make you a better photographer if you just stick to your old habits. It won't even make you a more versatile photographer unless you make sure you make use of your new camera's inherent versatility. But if you're trying to take pictures now that your old camera can't handle, a new camera can be the answer if you know what you want.

# It's a travelpod

**This compact cartop carrier changes like magic from streamlined cargo space to sleeping area to stable, shallow draft outboard**

**By CHARLES R. GRETZ**

IF THE CAMPING BUG hasn't had his stinger into you yet, watch out! Here's just the rig that could give you the fever. It's the perfect answer to family camping in a car. After an 8500-mile trip, I should know.

What's so great about it? Imagine being able to take off with a king-size, cartop luggage carrier that packs a whopping 36 cubic feet of waterproof tote space!

Imagine having an overnight sleeper when you arrive that's as easy to set up as an umbrella!

Imagine being able to take along your own boat for a day of trolling a quiet stream!

All this is possible with this all-in-one Travelpod that I designed, built, and tested with 300 pounds of camping gear. It turned out to be the most versatile and most functional camper on the road. What's more, we had the whole car to ourselves since everything can be stowed topside.

The Pod is designed to mount on regular steel luggage bars that rest on the wagontop and clip to the rain gutters. Access to the sleeper is by an aluminum boat ladder that stows on the underside of the sliding lid. To open the Pod on the road, you simply slide back its lid; it has handles to lift it off. Contents are safe from rain by virtue of a tongue-and-groove waterproof track and are secured from theft by foot-locker-type trunk latches and a padlock.

One person can set up the Pod in minutes for

SEE ALSO
**Boat camping . . . Boating, trailers . . . Camping . . . Cartopping, boat . . . Outboard motors . . . Recreational vehicles . . . Trailers, tent**

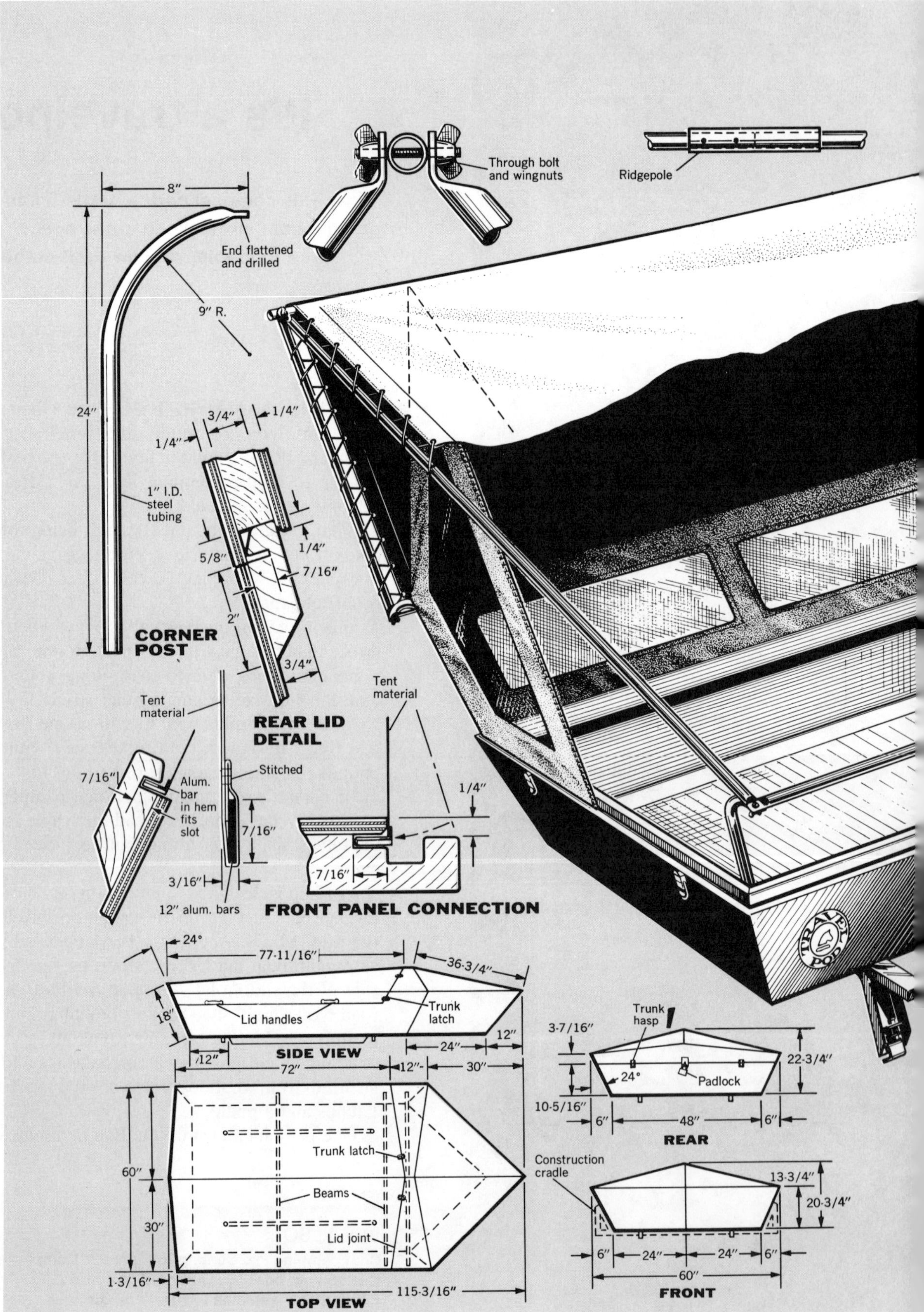
Through bolt and wingnuts
Ridgepole
8″
End flattened and drilled
9″ R.
24″
1″ I.D. steel tubing
CORNER POST
3/4″
1/4″
1/4″
1/4″
5/8″
7/16″
2″
3/4″
REAR LID DETAIL
Tent material
7/16″
Alum. bar in hem fits slot
Stitched
7/16″
3/16″
12″ alum. bars
Tent material
1/4″
7/16″
FRONT PANEL CONNECTION
24°
77-11/16″
36-3/4″
18″
Lid handles
Trunk latch
12″
24″
12″
SIDE VIEW
72″
12″
30″
60″
30″
Trunk latch
Beams
Lid joint
1-3/16″
115-3/16″
TOP VIEW
Trunk hasp
3-7/16″
22-3/4″
24°
Padlock
10-5/16″
6″
48″
6″
REAR
Construction cradle
13-3/4″
20-3/4″
6″
24″
24″
6″
60″
FRONT

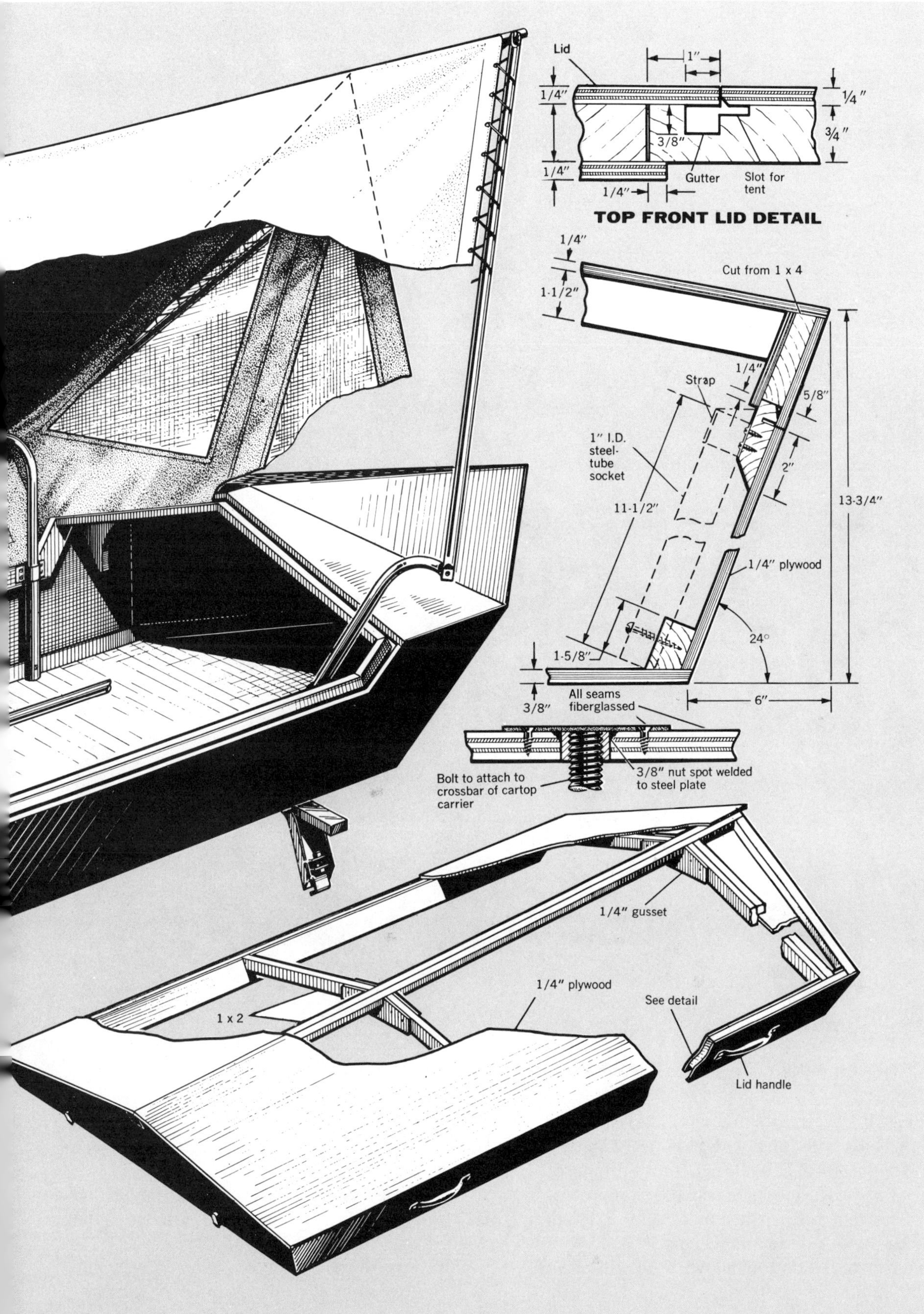
Lid
1"
1/4"
1/4"
3/4"
3/8"
1/4"
Gutter
Slot for tent
1/4"
TOP FRONT LID DETAIL
1/4"
1-1/2"
Cut from 1 x 4
1/4"
Strap
5/8"
1" I.D. steel-tube socket
2"
11-1/2"
13-3/4"
1/4" plywood
24°
1-5/8"
3/8"
All seams fiberglassed
6"
Bolt to attach to crossbar of cartop carrier
3/8" nut spot welded to steel plate
1/4" gusset
1/4" plywood
1 x 2
See detail
Lid handle

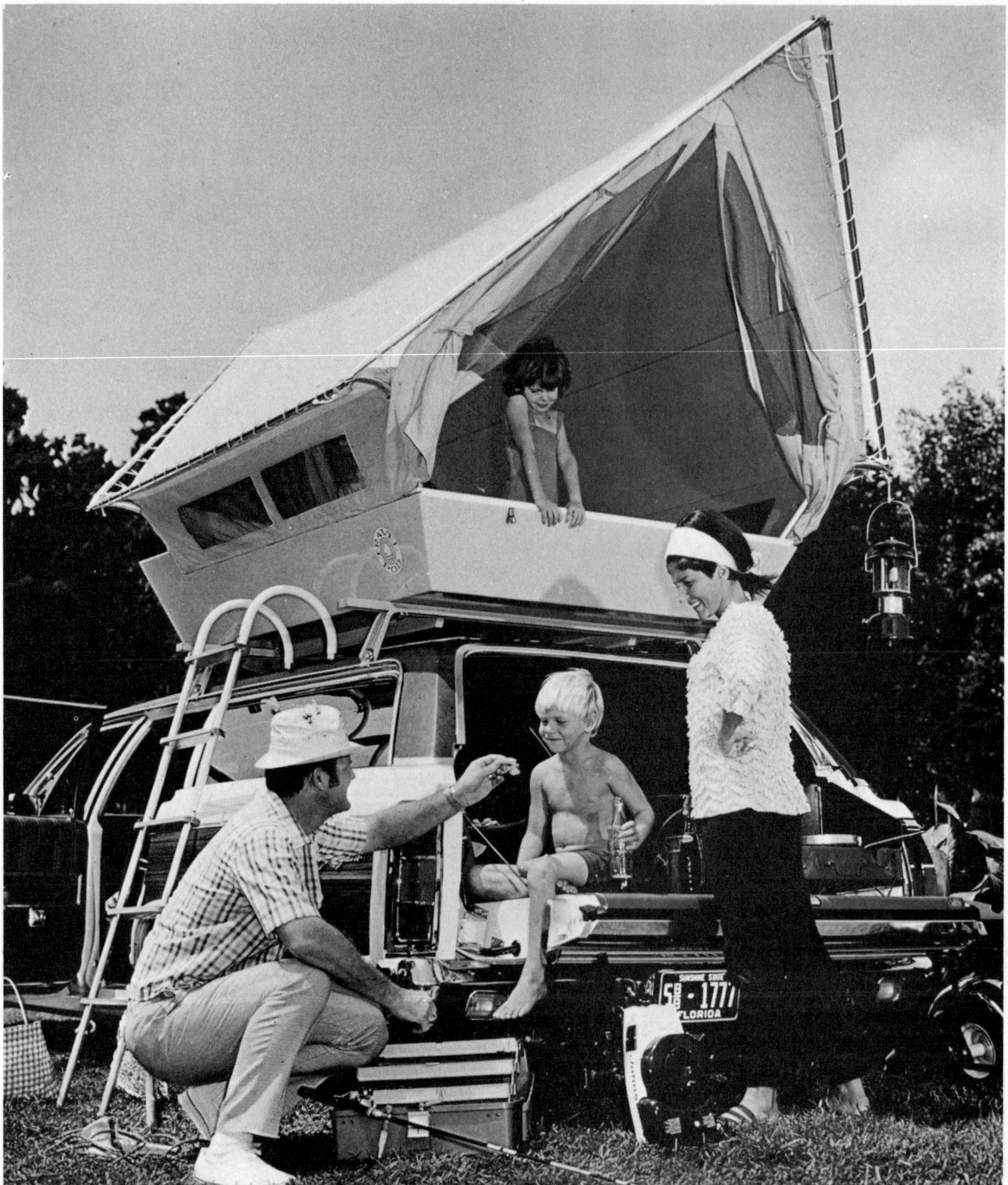

**SPACE FOR THE FAMILY,** along with a boat and spacious cargo volume, are part of Travelpod.

sleeping. It's draftproof and you'll have no fear that the kids will fall out. Nylon-net windows ventilate the tent and, by design, the overhang of the roof provides adequate sun and rain protection. The tent rolls in a bundle only 8 in. in diameter by 6 ft. long and weighs 15 lbs.

The broad, flat bottom allows the Pod to be a stable, minimum-draft boat. A small electric trolling motor is ideal to propel it and will run all day on a fully charged car battery. There's plenty of shade under the nose cone for an ice cooler, even a mini-TV set.

On the road, the Pod is less than 2 ft. high and is designed to present minimum resistance to the wind. Yet it pops into instant service.

For the most part, the Pod and its lid are made of ¼-in. marine plywood. Members that form the tongue-and-groove tracks are of solid stock.

**WHEN BUSINESS** requires long trips at night, bright quartz lights aid safety. Legality of such lights varies from state to state.

**INDOOR-OUTDOOR COOKING** is easy with this door-mounted ice chest and hook-on galley box. They're attached under the awning.

# Camp/work conversion for a van

By JIM ELDER

■ SOMEWHERE between Detroit and California, the van has evolved into something more than a delivery truck. Some of us must make our vehicles work *and* play, carrying passengers, cargo and tools during the week, then kids, bikes, boats and tackle on weekends. A "camp-and-work bin" van conversion seems a practical compromise. As our basic vehicle we chose a bare Ford "Display Van."

We divided our add-ons into three categories:

■ Safety—quartz lights, fire extinguisher, grille

SEE ALSO
**Trailers, tent . . . Vacation homes**

**EVEN A HOBIE CAT** can cartop aboard this van, and a step ladder hinge-mounts on the van roof without straining the door.

**WITHIN EASY REACH** of the driver are rucksack on the door, a CB, a kit bag Velcro-attached to the dash top, and a serving tray.

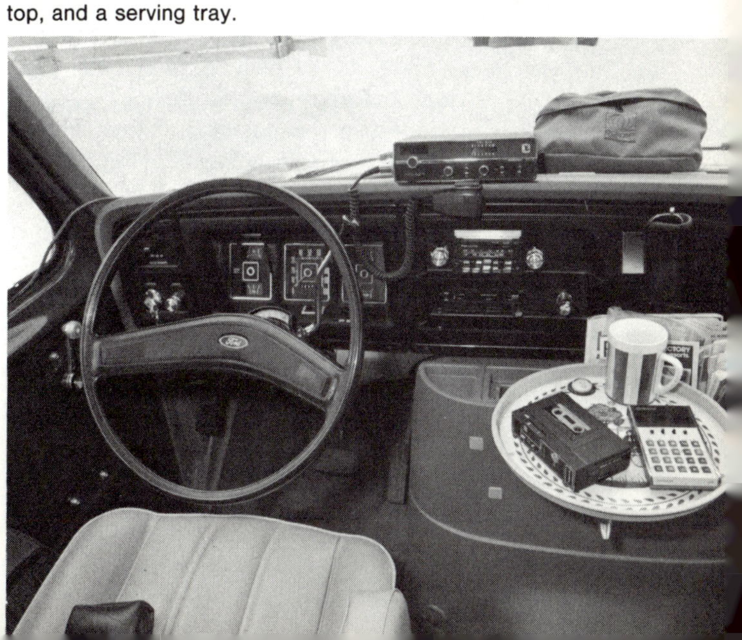

**PLANS ARE** checked before tracings for panel cutting are made.

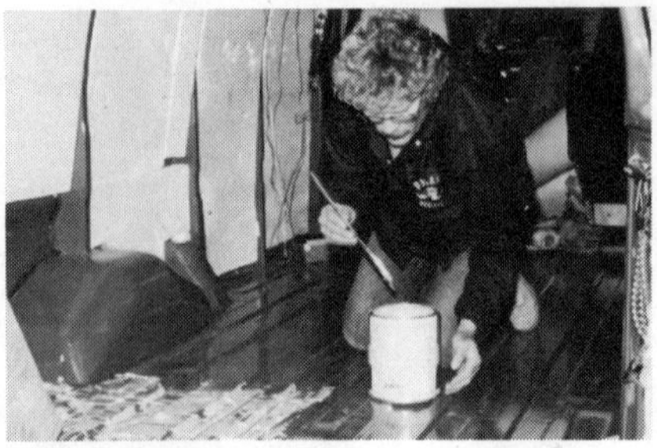

**HOLES FOR** floor screws could start rust, so Ensolite was cemented in.

**EXTRA HANDS** help bow ceiling into place. They are a jack, bat, tripod.

**VIEW AFT** shows ice chest and stove (left), rear seat with speakers and flag.

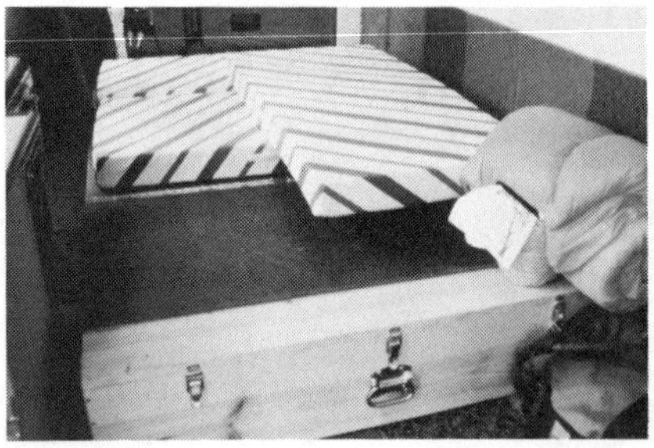

**STORAGE** boxes are in place and cover unfolded. Cushions become mattress.

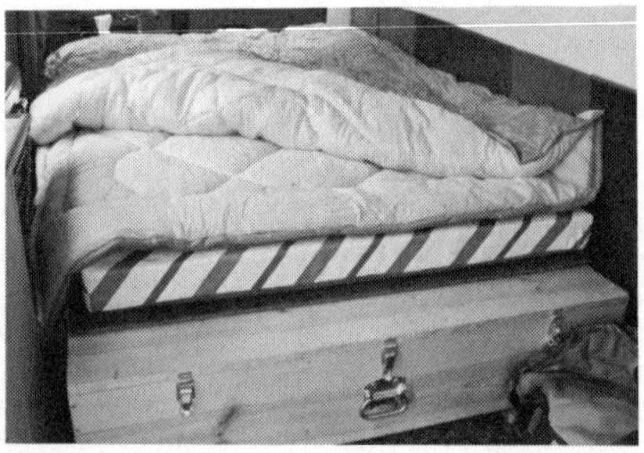

**SLEEPING BAGS** complete bed made from boxes, awning-covered cushions.

guard, special tires, alarm system, CB radio.

■ Utility—hitch, top rack, cargo capabilities.

■ Comfort and convenience—airconditioning, seat/bed system, paneling, insulation, interior lighting, roof ladder, radio/tape deck, storage, cooking and refrigeration, awning and many more luxuries.

Good sense suggested doing all the safety items first, but because we would be spending hours working in and around our Ford van we started with a sound system. For half the factory installation cost, you can buy dozens of aftermarket radio/tape systems.

### inexpensive extras

After some luxury electronics, it was time for pinchpenny projects. A Coleman rucksack on the driver's door and a zipper "fannypack" on the passenger door helped end cockpit clutter. From the hardware store came holders for flashlight, tire gauge and ice scraper. From the boat stored for the winter we pirated a fire extinguisher and cigaret lighter with map light. (When we removed six screws and pulled the dash panel to install the radio, we found Ford has a 12-volt wire waiting for do-it-yourselfers.)

Midwinter was no time to run around with bare van walls, but with no immediate money or time for a full interior paneling job, we glued in 2-inch Styrofoam panels and filled the odd-shaped spaces with aerosol urethane foam. Fiberglass in the roof was held temporarily by the factory-equipment cardboard panels. Interior temperature now stayed up and the noise level went down. We moved the spare tire outside on a Dick Cepek rear mount. Also from Cepek: a Filtron air cleaner, dipstick wiper, plus interior lights.

With 2000 miles on the odometer, we drained the break-in lubricant, flushed the engine and dumped in six quarts of Mobil 1 synthetic along with a new filter. Highway mpg jumped from 15 to 17 and cold winter starts sounded like summer.

### triple-duty storage boxes

Saving money, we built three plywood boxes, 24 x 24 x 12 inches, and one larger 48 x 12 x 12-inch box. The long one holds tripods, reflectors and fishing rods. The squares carry sleeping bags, food and camping gear. Pushed back they are seats and spread out they support a hinged bed—three 2 x 4-foot Masonite panels which are topped by three cushions. Bedspread and rainfly covers, home-sewn over 4-inch foam, serve as seat cushions or a mattress.

Night driving can be a nightmare if you're used to European quartz lights. We next installed Marchal H-4 headlight conversions, legal in Oregon and Washington, tolerated by intelligent Bears in most other states and illegal in California. On an Off-Road Performance grille guard, we mounted Cibié fog and driving lights. If you think $150 worth of quartz-halogen lighting and $100 more in a grille guard violates the "budget" approach, just drive once with good night sight and price the bodywork bill for even a minor front-end dent.

After 12,000 miles, summer finally came, and to cool the van we installed a Coleman Polar Cub II evaporative airconditioner. It uses water and 12-volt power, works well in areas where the relative humidity is low to moderate, doubles as an exhaust fan, and costs less than half the factory-air price. It can be cut in and hooked up in one afternoon, and we mounted it on the left side of the roof so we could still cartop our noncar cool-offs—canoes and Hobie Cats. A Quick-N-Easy roof carrier anchors our homemade awning, also created from a tent rainfly, and transports canoes. A T.H.E. rack cartops the 16-foot Hobie Cat.

To get up top, we needed a ladder, but most van ladders bolt to a rear door and Detroit didn't make those hinges to support heavy climbers. Our selection, Bill Devin's Safety Step ladder, mounts on the roof, rests on the bumper, and swings up to allow opening the door.

Many things we work or play with are too big to ride on top—snowmobiles, larger boats, trailers full of firewood, antique Model A hulks we find and tow home to restore. From Valley Tow-Rite we bought a Class III hitch platform. It took one hour to install, counting the holes Ford forgot to drill. They didn't forget the wiring harness, however, so hooking up the trailer light connections took only 10 minutes. We used a Bargman seven-wire heavy-duty socket so we can plug into anything the Ford can pull.

**swing-out 'kitchen'**

Next, a side door was modified to swing out with a Coleman upright cooler. We bolted this ice chest to the rear side door and then built a cook center to fit over it. This contraption holds a Coleman LP-gas cookstove, provdes a work surface, and features a combination storage/service shelf which looks like a metal dairy delivery basket. That's because it is, but it wasn't ripped off. We asked the dairy if we could buy one, and they were so overwhelmed by the honest offer that they gave it to us.

In an on-the-go kitchen, stoves rattle and all pans have hot handles. Pack a hot mitt under the stove lid so you have it along and it will also stop the metal noises. Mount the cooler so that when the van door holding it is closed, the cooler rests on the van floor.

After 15,000 miles, we finally got around to paneling. T&H Van Works Unltd., Canoga Park, CA., understands that not everyone wants a fur-lined surfer van (although they supply all those customizing accessories, too). We ordered a set of full-size cutting plans from them. Home carpenters should check every dimension and rib spacing since all vans are not assembled just the same. Our Ford proved close to the T&H patterns, and we used ¼-inch A-C fir plywood since we wanted something cheap and tough, with enough substance to take screws. We covered it halfway up with carpet sample squares that are gougeproof, easy to clean and sound-deadening. Above that we glued on inexpensive lightweight canvas sprayed with a stain repellent for easy cleaning. On the ceiling are the maps and flags that are personal favorites of ours when it comes to decorating.

**after 20,000 miles, success**

After 20,000 miles and almost a year into our project we can report the van a success. Seat-bed boxes and camping supplies are pulled out easily to make room for equipment required on business trips. Smaller refinement projects are still underway, with many more planned. An old tripod swivel head and quick clamps were adapted to a serving tray and a clipboard. The tripod head is attached to the engine cover next to the driver's seat. Notes are made on the clipboard, or the tray is quickly mounted instead to hold a pocket tape recorder or coffee cup and snacks.

Projects we'll be starting soon include courtesy-light switches on all doors and a security alarm system (we strung the wires for both before we paneled). We also want to add a hanging closet, arrangements for hanging wet wetsuits, life jackets and flyfishing waders as well. A bracket for a small outboard motor will be useful plus an outside bracket for a jerrycan of outboard fuel mix. A roof-mount Marchal quartz-halogen spotlight is on our future list, and a dash mount for a Bearcat 210 Scanner. Boman's tempered-glass roof hatch is built and priced for yachts, but is our choice for a great van skylight. And for cold winters we may install a Coleman LP-gas furnace. That should keep us fairly satisfied for nine more years.

You don't think our van will last that long? Ours will. We bought the Ford Shop Manual, Service Specs, and Wiring Diagrams. That $25 has already paid off. The van has never been back to the shop; we do our own preventive maintenance. If trucking firms can get a million miles from their rigs, we can hope for our yearly 50,000 times 10.

**This simple box cover bolts over your truck bed and turns it into a snug shelter cabin. If you want more headroom, just add the high-rise sides**

By RAY SHOBERG

# Hi-lo camper for your pickup

SEE ALSO
**Backpacking . . . Boat camping . . . Camping . . . Hunting . . . Recreational vehicles . . . Trailers, tent . . . Vacation homes**

PLASTIC WELT called "gymp" is stapled along the inside edges of the panels before they are assembled as shown in the photo above. Like quarter round, this covers the joint once the pieces are assembled. Plywood gussets on the side panels, right photo, give curve to the roofline and anchor the brackets which support a pair of carrier bars as sturdy as necessary.

DON'T CONFUSE a camper shell with a standard pickup camper. The shell bolts directly to the top flange of your pickup box, and unlike a camper, it has no lower sidewalls or floor. Ultralight, it can usually be lifted on or off the truck without the aid of winches or jacks.

The two-part shell detailed here has several advantages over commercial shells. When you buy a shell, you have to choose between a low-profile model roughly the same height as your truck cab and a high-rise style which offers standup headroom.

The higher shells are more comfortable to live in, but they are also bulkier and offer greater wind resistance. Low shells are considerably more cramped. However, they're more compact, lighter and generally handier for year-round utility shelter for your pickup box.

With this two-part shell, you can choose the height to suit the job. The shell itself is a standard low-ceiling design. However, sandwich in the extender between the shell and truck box and you have a stand-up shell. You can also use the extender alone to increase the height of your pickup box for hauling bulky loads. Temporary hooks screwed into the top plate will provide anchorage to lash down the load or secure a tarp.

Before starting construction, it's a good idea to have all the materials and fittings on hand so they can be measured exactly. Window openings, for example, shouldn't be cut until you have the windows available to use as a cutting pattern. This way you'll be assured of a perfect fit.

You may have a problem finding some of these parts through local sources. If so, you can order the windows, aluminum skin, metal roof and moldings from camping supply dealers.

From the drawings, you should have no trouble visualizing the general construction of the shell and extender. Both are framed with 2x2s and faced with aluminum skin on the outside and plywood paneling on the inside. Naturally, dimensions must be tailored to suit your pickup.

While you can leave the walls hollow, it costs very little to insulate them with batts of fiberglass. This will make the shell cooler in summer and warmer in cold weather. It's well worth the slight extra cost.

If you want to work indoors, the sides, front and back can be partially finished in your workshop and then assembled later outdoors. Construct the framing and attach the inside paneling. Before doing the roof, however, check the dimensions against those of your shop door to make sure you'll be able to get it out.

Take all length and width measurements from the pickup box itself. Height is a matter of taste, but remember that the roof should protrude slightly above the cab if you plan to haul loads on

**FIBERGLASS INSULATION** is well worth the slight added expense. Note that the front 2x2 has been rounded off to conform to the shape of the plywood gusset.

**ROOF VENT FRAMING** should be installed now, even though you may not put one in immediately. Drill tiny locator holes in the corners for cutting opening later.

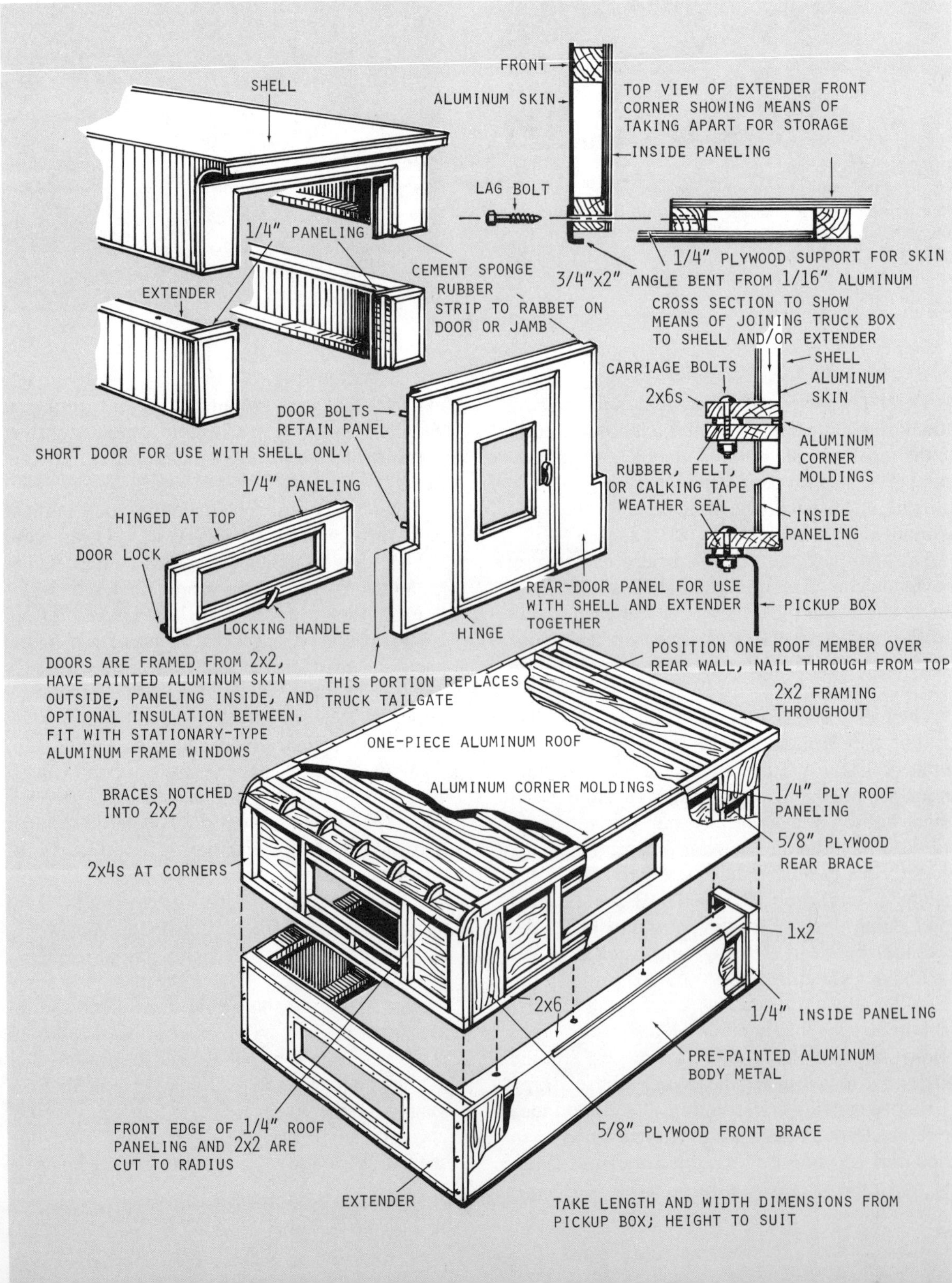

# Bring your camping first aid up-to-date

By WILLIAM W. CONNER, M.D.

VACATIONERS ARE FREQUENTLY far from medical help when an injury occurs—so it's important to know alternate first-aid methods for some common ills:

***Fishhook penetration***: The old recommended method for removing a fishhook was to force the point out through the skin and cut off the eye so it would pass through the hole thus made.

An equally good, and often less damaging to the victim, method employs an 18-inch length of string which is looped around the bend of the hook and held slack. With your other hand grasp the eye of the hook and press the shank down in the direction of the point. At the same time the shank is depressed, give the string a sharp jerk—and out snaps the hook. A little disinfectant over the wound, and you're ready to bait up again.

***Snake bites***: America has only four types of truly venomous snakes: rattlesnake, copperhead, moccasin and coral snake. Other snakes may bite when alarmed, but they lack venom and fangs.

A new method used by doctors in handling snake bites is to remove a patch of skin from the site of the bite about the size of a half dollar, along with the underlying tissue. The venom usually collects here in large quantities, and removing the tissue rids the victim of most danger. The limb is then treated with antivenom injected into an artery, and a new piece of skin is later placed over the wound. Most of this is obviously best left to the doctor. What, then, can be done in the way of first aid?

First, allow the victim to expend as little energy as possible. This helps slow down circulation and delays absorption into the system. Second, apply a snug tourniquet about two inches above the bite. That is, between the bite and the heart. It is a good rule of thumb that if you cannot easily push your finger under the tourniquet, it is too tight. A greater danger results from applying it too tightly than too loosely; a leg can be lost if the blood supply is cut off for too long. Third, transport the victim to a hospital or a doctor.

Note that no mention is made of cutting or sucking on the bite area! However, if the accident occurs where it may take many hours to reach a doctor, and the bite was from a moccasin, a snake whose venom is particularly destructive to injected tissue, your first aid should be more extensive. The area of the bite should be cut out. All that is necessary is a clean, sharp knife. The venom itself provides good anesthesia. Make a circular incision about two inches in diameter with the fang marks in the center. Cut deep enough to include the fat that lies under the skin. The venom may have blackened the fat, and by picking out all of the blackened tissue visible, you can remove much of the stored venom. A careful cleaning of the wound with a good disinfectant before and after the operation and a sterile or clean pressure-dressing will suffice to ward off infection until you get medical help.

More complications can arise from an infected wound than from the snake bite itself! So avoid making any cuts in the skin unless a long delay in reaching the doctor is unavoidable.

***Bleeding***: Bleeding from any cause can usually be treated most effectively by applying firm pressure with a clean cloth directly on the wound. Forget tourniquets and pressure points unless the bleeding is from a severed limb.

***Burns***: Perhaps the most common vacation mishap is a burn—sunburn, campfire burn, lantern burn and others. The treatment is similar for all. The prime consideration is to stop tissue damage as quickly as possible, and the best counteractant for heat is cold. So plunge the burned area into cold water immediately—a cold stream, a pail of drinking water, or over the side of a boat. Hold it in the water until the pain subsides. Many otherwise serious burns treated in this simple manner can be reduced to not much more than a moderate sunburn. A clean and dry dressing or a small amount of antibiotic ointment can then be used to cover the area. Avoid the age-old "goo treatment" or butter, baking soda and the like. These applications often make matters worse.

# Sew your own camping gear

**Several hours of easy work can cut your costs in half and at the same time assure you of quality equipment which will last through many camping trips**

**By BILL McKEOWN**

SEW-IT-YOURSELF Aspen jacket comes in kit from Frostline for about $26. A few hours' work converts pieces into a quality down-filled coat valued at $50.

HOW WOULD YOU like a 50-percent discount on camping gear that you need? Not cheap equipment, either, but the very best obtainable—finely made and probably better than any you could order anywhere.

Nor is your choice limited to a handwoven belt or watch band. Look over the list below and see if there isn't an item—or, more likely, many—that you wish you had:

Down-filled sleeping bag
Bag liner
Bivouac cover
Down jacket
Parka
Ski coat
Down vest
Down sweater
Down pants
Gaiters
Poncho
Booties
Overboots
Chaps
Rain gear
Tent
Stuff sack and backpack
Bike bag

Many of these come in assorted sizes, types and colors, while the variety of items offered is growing all the time.

At first glance, a price tag of $100 for a sleeping-bag kit or tent kit can seem steep, especially since you still have to put it together. But if you don't enjoy working with your hands, you probably wouldn't take up camping, and the kit manufacturers have found their customers already know about but don't want the flimsy cut-rate items from some surplus outlets. For example, if you buy a sleeping-bag kit for about $84 which is rated for −20° F. and then add the optional $21 worth of "expedition down fill" for even more insulation, it's obvious you want the best. Or perhaps you select a model which weighs about five and a half pounds for about $60. You can sew the special full-length foam pad right into the down bag, where it is always ready for use.

Because you or your wife do the sewing, you can be sure that no one took shortcuts on inside seams that don't show, and you can also double stitch or lock stitch spots where you know from experience a zipper can tear or a D-ring pull loose when you're miles from nowhere. Basic sewing experience is helpful, though not essential, and these kits could be sewn by hand rather than machine if you had a lot of time. According to a spokesman for the EMSkit division of Eastern Mountain Sports, "Novices seem to understand the instructions more easily than experienced seamstresses, and the most satisfied customers are those who enjoy making their own equipment rather than those who try to save money." And any man who thinks that only women can sew has never been a bachelor very long, or hasn't visited a sail loft recently.

To test kit instructions and materials, and because they looked great, we first selected down booties, about $6, and overboots, about $5, avail-

SEE ALSO
**Backpacking . . . Bicycles . . . Binoculars . . . Boat camping . . . Campers . . . Canoes . . . Emergencies . . . Hunting . . . Map reading . . . Trailers, tent**

able in blue, green, or orange. Both came with all the materials—thread, elastic, cord and laces—plus illustrated instructions so explicit that someone who had never sewn before could have worked them out.

Even after years of sewing, Virginia Hricko had never edge-seared fabric, but the instructions made it easy. You carefully pass a flame along the edge of the Ripstop Nylon without touching it. The heat seals the edge so that it will never unravel during wear, washing, or cleaning. It is easy to see why in normal factory assembly there's not time to spare for refinements such as this.

Many of the products come in a variety of sizes, ranging down to clothing for children and pack carriers for babies. Some of the sleeping bags start with 42-inch lengths, and then can later be extended with 12-inch sections as the youngster grows. Provided in small plastic packs is the down you will use to fill the insulated clothing and bags. This, we found, is a neat and easy way to handle the very fluffy feathers.

And even if you have no plans for camping or cool-weather sports, the kit makers' catalogs deserve study for their versatile bargain travel packs and bags. We test-constructed kits from each, and found them uniformly well designed. They were complete with all necessary parts, and with instructions that told both the how, and reassuringly, the why of each succeeding step.

Cross-country and downhill skiers, hunters, bicyclists and travelers using almost any other means of getting around are likely to find kits designed especially for them.

Though it would be hard to select our favorite kit to sew up and use, our best bag-title award would certainly go to Eastern Mountain Sports for their $16 model in navy or orange. It's the EMSkit Knapsack-snackpack Rucksack. It proved to be easier to sew than to say.

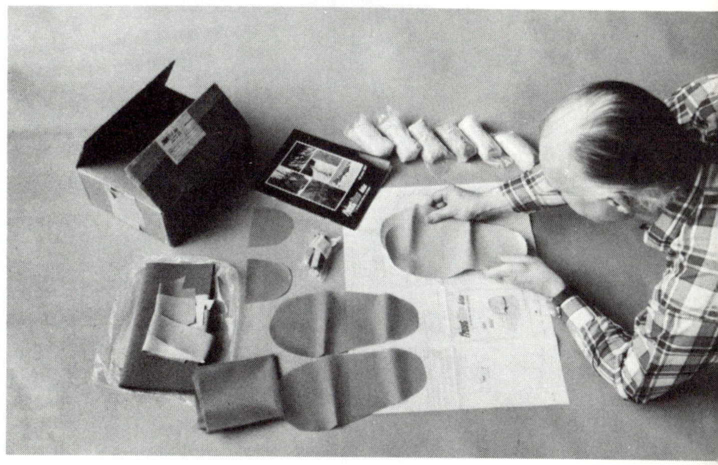

**TURNING PARTS** into a pack took an evening, Virginia Hricko (top) found, after following kit's complete instructions and using all the comprehensive enclosures. Fabrics, goosedown packs and thread were in the boot and bootie kits (center). Final reward of needlework (below) is a bargain in quality gear.

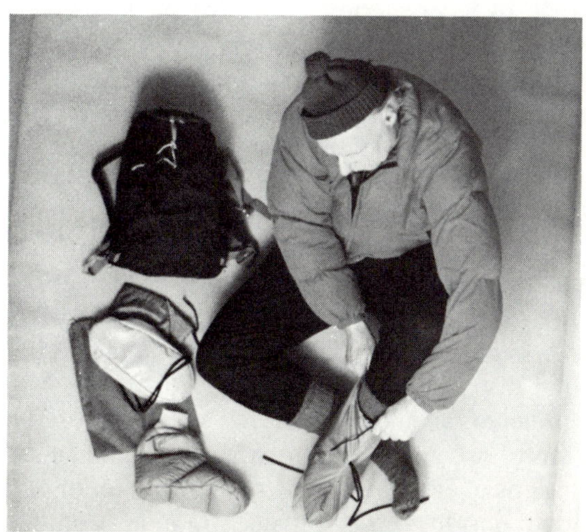

**Kit manufacturers**

**CARIKITS**
Holubar, Box 7, Boulder, Colo. 80302.

**EMSKITS**
Eastern Mountain Sports, Inc., 1041 Commonwealth Ave., Boston, Mass. 02215.

**FROSTLINE**
Frostline Kits, 452 Burbank, Broomfield, Colo. 80020.

**MAKIT**
Mountain Adventure Kits, Box 571, Whittier, Calif. 90608.

# How to pick a pack

By RICHARD DUNLOP

■ YOU DON'T HAVE to be an equipment freak to be concerned about the pack for your back. The wrong one can yank you backward just when you want to lean forward into a hill to let gravity help you on a steep climb, and it can catch in the brush and hang you up. A poorly fitted pack can give you more varieties of backaches than you ever imagined existed, and make you wish you'd stayed home. On the other hand, you don't have to take a money's-no-object approach to achieve backpacking comfort. The idea is to match the kind of pack you use to the kind of backpacking you plan to do.

There is always controversy over the pros and cons of soft packs, packs with frames, and which designs of each are best. You'll hear a lot of talk about inside versus outside framing, contour frames and flexible frames. Almost every one of the nation's 5-million backpacking enthusiasts seems to have his own surefire suspension theory. A little common sense discussion of the facts may help.

**the ancient art of backpacking**

From the first North American use of a pack frame by Sioux and Cheyenne Indians—even back to the vineyard workers of Europe carrying baskets of grapes—it was obvious that the frame distributed weight more evenly across the back and shoulders. As Mic Mead, president of Adventure Pack at El Cajon, CA, explains, "It provides fixed points to attach the load to the wearer." Eventually recreational hikers took up the pack frame of the mountain men. As long ago as 1878, Henry Merriam constructed a pack and frame with good hip-suspension.

In 1922, Trapper Nelson, a Seattle hiker, began to manufacture a pack frame made of spruce covered with a tightly laced piece of canvas to act as a back pad. This frame, and countless frames copied from it during the twenties and thirties were popular until immediately after World War II when A. I. Kelty invented the arc-welded contoured aluminum pack frame with padded shoulder straps. To go with the frame, Kelty created a nylon pack with easy-to-get-into pockets, compartment design allowing for variable load placement and a waist that would help distribute weight more evenly on the hips. A Kelty, or a not-too-different Camp Trails, frame pack is lightweight and allows a load to be carried high and close to a backpacker's center of gravity so that he doesn't have to lean forward like a Neanderthal when he walks. These packs soon became the most popular designs used on North American trails, and they remain among the favorites today.

**stay loose, says one designer**

"Modern pack designers, however, always overlooked one critical factor," says Murray Peltz, director of research and development at JanSport in Everett, WA. "The human body does not remain rigid while walking. Merriam had made this simple observation when he designed that 1878 pack so that the bag and frame could move independently on the hiker's body."

One answer to the rigidity problem was the soft pack. The Germans invented the rucksack. A good rucksack fits the body well and moves with it. Eventually it replaced the American knapsack because it was suspended from the strongest part of the shoulders and thereby put less strain on the back. With most of the load on the shoulders, nevertheless, this meant muscle fatigue not only in the back but in the legs as well.

Today's rucksack is fashioned of tight canvas to make it light in weight, but since it is only a sack with a draw cord to bring the top together, its interior is usually a jumble of articles. And since it is close-fitting, it also is often too warm on the back during a hot day's hike. Pack it properly, or sharp and hard objects inside can poke and rub against your back.

**frame up**

A pack frame makes it possible to ignore placement of objects in the pack except for the usual necessity of placing heavy objects high and close to the body. Most backpackers today find the frame pack invaluable for long-distance hikes because it keeps its shape well and is easy to live out of. Not everyone agrees about its superiority, however, because soft packs are now available that control the distribution of the weight through compartmentalization. Lawrence Horton of Rivendell Mountain Works, Victor, ID, is an advocate of the soft pack, and is blunt in his criticism of the frame pack:

"The idea of a frame was originated for the purpose of carrying heavy, awkward-shaped objects too hard, sharp or bulky to carry against one's back. They are ideally suited for carrying giant loads—75 pounds and up. In fact, if one is not carrying at least this much weight, or something like oxygen tanks, gasoline drums, or maybe pig iron, the advantages of a frame pack are questionable. They are bulky, cumbersome, heavy, fragile and expensive. Current designs, loaded with gadgetry, tend to become more and more cumbersome."

**inside job**

To Horton, it appears that the backpacking public has become so conditioned to the necessity of a frame that "when confronted with a soft pack, they have no idea where to put their sleeping bag—no frame extends below the pack!"

The soft pack began its present popularity rise in the early 1960s when Don Jensen, a Harvard student with a penchant for climbing Alaska's peaks, worked out a design for a pack that transfers much of the load directly to the hips. It is a compartmentalized shaped pack which fits snugly into the contours of the back. Its adherents say that its weight seems to become part of the hiker's body and that the pack does not move or sway. On the negative side, because it hugs the body so closely, it can be hot on a torrid day.

It is limp when unloaded and gains vertical shape only when it is stuffed. Many long-distance backpackers say it can be hard to live out of. Whatever its drawbacks, however, it has been widely imitated. Rivendell Mountain Works makes the original in four lengths to insure proper fit to a hiker's back. The first Jensen pack was built to carry about 50 pounds comfortably, and held approximatley 3100 cu. in. Now there is also a Giant Jensen with 4400-cu.-in. capacity, designed to carry upward of 75 pounds.

The internal frame pack is also popular with many, particularly for relatively short backpacks. It does not hug the back as well as the Jensen soft pack, but it keeps vertical shape better when it is empty.

**adjustable fit**

Recently there has been a strong trend toward pack frames that employ adjustable fittings on the shoulder level cross bar instead of arc-welds in order to make the frame more flexible. JanSport, a leader in the manufacture of frame packs employing this kind of frame, and the firm that first introduced it in 1967, claims that the "mechanically secured joints allow the frame to flex with the body's motion while hiking." High Touring, the Salt Lake City backpack manufacturer, also makes its frames adjustable.

"If the trail is long, points of wear on your body must be constantly changed to prevent sore spots. The more points of adjustment, the better," comments Frank Cunningham Jr., of High Touring.

Lowe Alpine Systems of Boulder, CO, also makes a flexible frame that is winning much

**PACKS FOR** every purpose range from daypack rucksack (left) and soft-stuff sack pack (center) to a full frame pack (right). Three shown here are sew-it-yourself pack kits which are available from Frostline

favor from veteran backpackers. LAS claims that "for the average backpacker who sticks only to established trails, the rigid design is fine. But for the advanced hiker, the extreme lack of versatility and flexibility of a rigid frame may outweigh its few advantages."

When it comes to strength, JanSport refers to multicycle torque tests performed by Ron Barstad and Dr. Stanley Chen of the Engineering Department, Arizona State University, that indicate their frame had not failed when the test was halted at 500 cycles, while the next strongest frame failed at 350 cycles. The Alpine Designs Adjustable Frame was also tested at Arizona, and was the only frame rated superior in quality of material and construction, excellent-superior in design and comfort. Alpine Designs in Boulder also reports that its new mini Flexpack is so adjustable that it can perform as an adult ski touring pack during the winter and a child's frame pack during the summer. To would-be customers for the full-sized Flexpack, Alpine Designs says, "Lift the frame over your head and crash it to the floor as hard as you can."

### picking the perfect one

With something like 35 United States manufacturers offering better than 300 kinds of backpacks, selecting a pack for your own use can be mystifying. At least it can be said that the frames of reliable manufacturers are all reasonably well made. The weight range of the frame is not too important. Heliarc welding is considered the strongest, but eutectic welding holds up just about as well. Aluminum fittings are satisfactory, too.

Frame contour has been found to be important. Let the frame bend forward at the top and it moves the center of gravity forward. The hold-open-bar carries much of the weight of the load and should be attached to the frame and not, as in some of the inexpensive models, used in a fabric pocket.

The belt is another key feature of a pack that should be checked. A good hip-suspension system distributes the weight evenly over as much of the hips as possible and keeps the load in close to your body. It should, at the same time, allow the back to adjust to the pack when making a steep climb or descent. This will reduce back and leg fatigue as well, while providing better balance and a feeling that the pack is secure. Some hikers claim that only the single piece belt permits the frame to ride smoothly on their hips. Others favor the two-piece belt.

### BACKPACK SOURCES

Among major companies in this field are the following. Some can supply names of dealers while others offer mail-order catalogs, services.

Adventure Pack<br>656 Front St.,<br>El Cajon,<br>Calif. 92020

Alpenlite<br>115 S. Spring St.,<br>Claremont,<br>Calif. 91711

Alpine Designs<br>6185 East Arapahoe,<br>Boulder,<br>Colo. 80303

Camp Trails<br>Box 14500,<br>Phoenix,<br>Ariz. 85063

Cannondale<br>37 Pulaski St.,<br>Stamford,<br>Conn. 06902

Coleman<br>250 N. St. Francis,<br>Wichita,<br>Kans. 67201

Frostline Kits<br>452 Burbank,<br>Broomfield,<br>Colo. 80020

Gerry<br>5450 N. Valley Hwy.,<br>Denver,<br>Colo. 80216

High Touring<br>1251 E. 2100 S.,<br>Salt Lake City,<br>Utah 84106

Holubar<br>Mountaineering<br>Box 7, Boulder,<br>Colo. 80302

JanSport<br>Paine Field Park,<br>Everett,<br>Wash. 98204

Kelty<br>1801 Victory Blvd.,<br>Glendale,<br>Calif. 91201

Lowe Alpine Systems<br>1752 N. 55th St.,<br>Boulder,<br>Colo. 80301

Recreational Equip.<br>Box 22090,<br>Seattle,<br>Wash. 98122

Revendell<br>Mountain Works<br>Box 198,<br>Victor, Idaho 83455

### choose quality

Examine not only the frame but the shoulder straps to see how well padded they are and whether they are wide enough so they won't bite into your muscles. Cheap ones skimp on pad material and sometimes use easily frayed nylon belting that can slip through buckle fasteners.

Stitching should be of durable nylon or cotton-wrap nylon, with stitches small, straight and even. Several companies provide do-it-yourself backpack kits at lower prices. Double stitching is required over pockets, and stress points must be reinforced. Zippers should be made of materials that will not jam in cold or damp weather.

There is a bewildering variety of pack compartments available. Most lines include a single-compartment model plus various dividers. In some you can unzip the floor of the upper section to make one big bag for large items. Others unzip around the outside for access to individual pockets. Somewhere there's a pack to best suit your purpose.

# Build a tin-can space heater

By ROY L. CLOUGH JR.

**Take the chill from your boat cabin, tent or small trailer with this inexpensive space heater. It burns Sterno and is made from household items**

SEE ALSO

**Backpacking . . . Bicycles . . . Binoculars . . .
Boat camping . . . Campers . . . Canoes . . .
Heaters, portable . . . Hunting . . . Map reading . . .
Recreational vehicles . . . Trailers, tent**

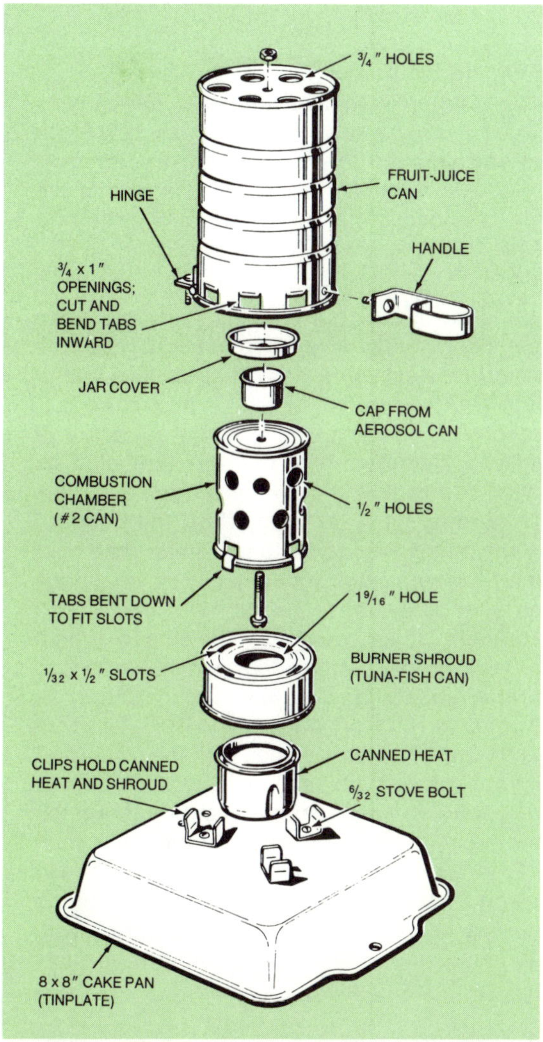


■ YOU'LL BE AMAZED at the blast of heat you get from a can of plain old Sterno when it is placed in this pint-size, inexpensive space heater. It is made from a collection of old tin cans and a pressed-steel cake pan which you probably have around the house. It is just the ticket for taking the chill from a boat cabin, tent or small trailer on those cold nights out in the woods.

The key features of the burner are a simple tuna-can shroud which insures a steady supply of air over the fuel to keep it from overheating and sputtering, and a burner pot which stabilizes the flame and insures that no unburned fumes escape. It's important that you follow the vent-hole sizes fairly closely. The cake-pan base never gets hot so it can be set down anywhere and you don't have to worry about it burning a tent or a table. One of the finer features is that the cake-pan base is also large enough that it is practically tip-proof.

The pull-apart drawing shows how the cans stack inside the hinged outer jacket which is a large fruit-juice can. The round holes can be neatly cut with a circle cutter, the square ones with an old wood chisel. In both cases, the metal should be backed up with a hardwood block to assure clean-cut holes. The clips on the base, plus the handle, are bent from short lengths of steel strapping.

**Pulling the pigeonholes**

*I want to refinish a shabby older-type desk in which the pigeonholes and small drawer appear to be an inserted unit. I need to remove it in order to refinish the interior of the writing compartment. The unit won't pull or slide out from the front and I don't see any fasteners. Can you suggest a procedure?—James Coxie, Lansing, Mich.*

Such separate units are usually fastened with either screws or small wedges, rarely with glue. Check the back panel of the desk for tiny screws or brads, or wedges tapped in from the back. You should find one of the three types of fasteners, but be especially careful in removing them —otherwise there may be damage difficult to repair. After refinishing, be sure to replace the unit in its original position and reuse the same fasteners.

**Glue guidelines**

*I'm a beginning woodworker, having just recently assembled a workshop with a few power tools. I'm puzzled by the array of glues available for joining wood in cabinet work. Can you give me some guidelines?—Stanley Ball, Kansas City, Mo.*

For speed and convenience, many craftsmen use the familiar "white" glues, but these offer limited resistance to moisture. Other skilled workers prefer the epoxies, while some still use the older casein glues. The hide glues are less commonly used, though obtainable in refined forms. When waterproof joinery is called for, the experienced craftsman turns to the resorcinols, two-part plastic resin glues that, like the epoxies, are mixed at the time of use. All the adhesives I've mentioned are quite reliable if you are careful to follow in detail the instructions on the containers.

**Pesky spray cans**

*I use spray cans quite frequently, but even though I invert the can and spray out the residue as directed, the next time I pick up the can the spray tip is apparently clogged and won't work. What does one do to make these things work? —B.E., Wis.*

Look closely at the tip after spraying as directed; you'll often see a tiny bubble of the material covering the tip. This seeps back into the orifice and hardens. Thus the next time you try spraying you may get a fizz or two, a splatter of drops or, more likely, nothing. I've found that if I pull off the spray tip—after inverting and spraying out residue—and dunk it in lacquer thinner I have no further trouble.

**Waterproof 'shoes'**

*When we have an unusually heavy rain of 1½ to 2 in., I get about the same amount of water in my basement. I'm new here and don't know the reason for the seepage, but something must be done. I have furniture stored in the basement—no place else to put it—and most of these pieces have legs with casters. Is there some way to prevent water damage to the legs?—A.O., Tex.*

You're probably thinking of rusty casters rather than legs if you get only 1½ to 2 in. of water, although some legs may absorb water at the higher level. I know of one homeowner with a similar problem who keeps the legs "shod" with plastic sacks and those large fruit-juice cans (the 46-oz. size). He simply raises the pieces and slips the cans or sacks over the legs. This stunt offers protection unless the water should rise above the tops of the sacks or cans. Probably the sacks are best as they are obtainable in various lengths.

Appliances such as a wringer-type washer or the legs on workshop power tools or a workbench can be protected in the same manner.

**Salvaging old bench**

*I've bought an old, beat-up cabinet-maker's bench for my home workshop. I can repair the legs and rails but the top stumps me. It's good maple, 2½ in. thick but it's been hacked at the edge, chopped into and drilled through in several places along the length. Any suggestions?—P.G., Mich.*

Bought new, such a bench would cost you a pretty penny today, so perhaps it's well worth the salvage. At best the top is going to look rather patchy but with good repairs, it should be serviceable for years to come. If the holes are of standard diameters they can be eliminated by driving in dowels that have been dipped in glue, then cuttting them off flush at top and bottom.

Neat fitting inserts in those areas that have been "hacked" and "chopped" can challenge your skill as a craftsman but, with care and a razor-sharp chisel, you should be able to bring it off. Use a scriber to mark areas where waste is to be removed, scribing the lines rather deeply so the chips will lift along the scribed lines. Then cut straight down along the lines with the chisel and chip out the waste by making a series of repeat cuts equally spaced in the waste. Square up the bottom and sides of the notch at the edge, then cut inserts from maple for a snug, true fit. Spread glue on these and clamp them in place until the glue has hardened. Then sand or scrape (with cabinetmaker's scraper) to a smooth surface. Finish with paste wax.

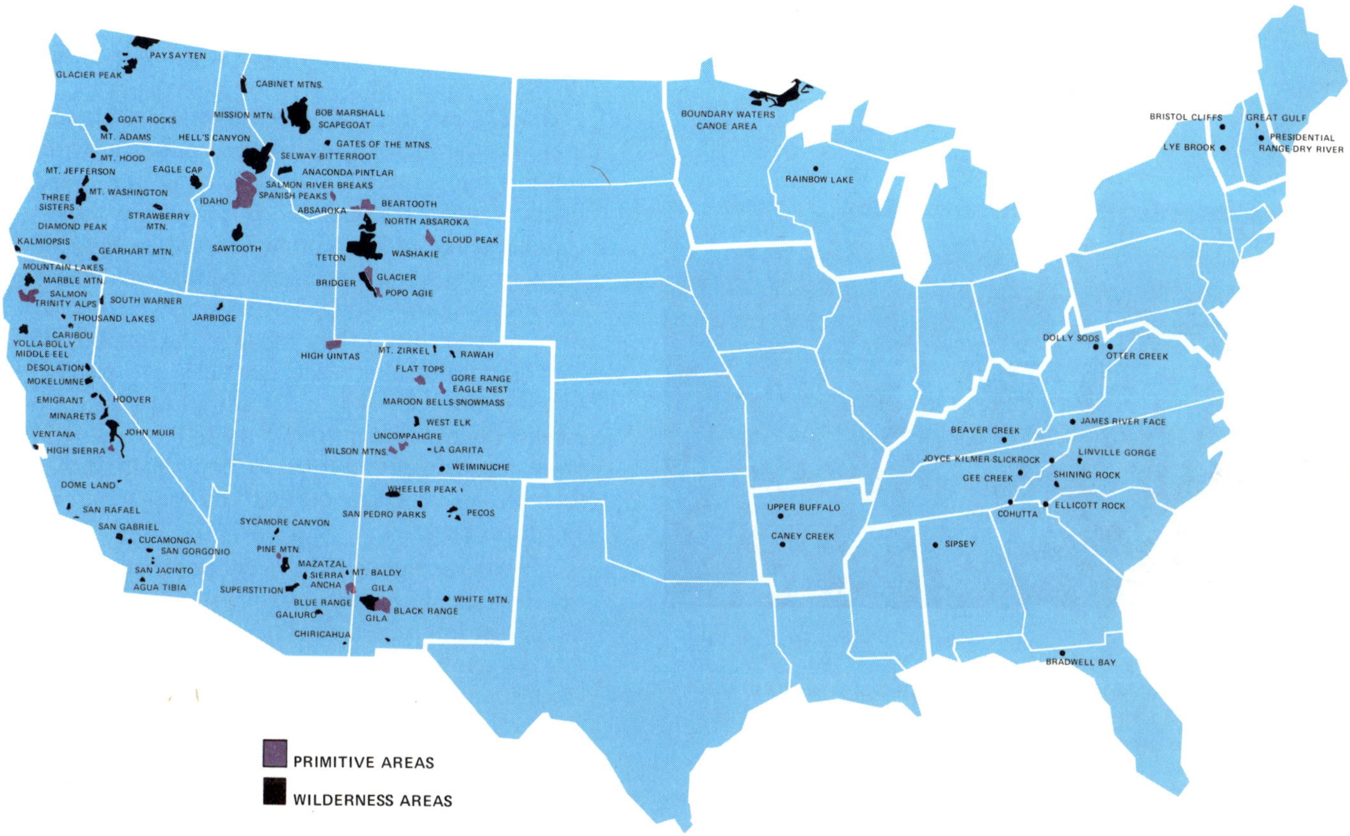


# How to camp away from campers

**Little-known, unspoiled wilderness areas have adventure-away-from-it-all appeal. Here are suggestions for places you can try and what you'll need**

**By JACK GALUB**

MANY SPORTSMEN consider wilderness camping the ultimate outdoor experience, although fortunately not enough people have tried it to spoil the sport. There are still a number of isolated areas where the stillness is broken only by the roar of a cascading stream, the snap of a branch, the rustle of game. For campers who look to the outdoors for solitude, these spots offer a special kind of escape.

When you go in, you go by foot, canoe or horse. No trail bikes, no four-wheel-drives, no runabouts, no recreational vehicles. You pack in what you need—there are no general stores, no trading posts, no souvenir stands, no lodges.

Camping the wilderness areas is adventurous but not dangerous if you are careful and well prepared. Before you start out, however, you should be certain that you know how to travel, cook and camp with light gear and minimum extras, particularly if you backpack; that you can navigate by map and compass, and that you are physically fit.

There are presently over 100 primitive and wilderness areas that have been set aside by the federal government to remain completely wild. These national forest categories are very similar;

SEE ALSO

**Backpacking . . . Bicycles . . . Binoculars . . .
Boat camping . . . Boating, white-water . . .
Campers . . . Canoes . . . Emergencies . . .
Exercising . . . Hunting . . . Map reading**

**REMOTE LAKES,** tall pine, silent crags provide true solitude, but they put camper to the test in the High Uintas Mountains of Utah. Altitudes from 8000 to 13,-499 feet and over 250 lakes for fishing are found in this area.

the primitive regions may eventually be incorporated into the wilderness system of the Department of Agriculture. The terrain of all these areas is so rugged that experienced wilderness men recommend you do not hike alone. A backpack party of at least three provides an extra measure of safety in case one member gets hurt, and also permits the dividing of community equipment and supplies into three or more packs. One hiker can carry the tent while others can share carrying the required food rations and cooking gear.

Map and compass can be keys to a safe and successful trip. While some main trails are marked, signs are kept to a minimum and there are none as you take side paths or move cross country. The Forest Service recommends that you obtain a map, study it, and then plot your progress as you go.

Finding your own way is part of the challenge and the satisfaction, but basic skills should be learned and practiced before you get into the wilderness. "We've had people lost and wandering around with their maps and compasses, not knowing how to use them," one ranger recalls. "We've also had others who just weren't in good enough shape after taking it easy all winter. They'd just fall apart on some of the mountain climbs and have to be carried down." While rangers will assist campers in primitive areas in an emergency, the general wilderness area attitude is, "You're grown up now, so use judgment. Don't take unnecessary chances."

Distances are deceptive out where the air is remarkably clear, and "nearby" streams, valleys and peaks can turn out to be miles away. Mountain weather can also worsen suddenly with freezing temperatures setting in at any time, particularly where the wind blows down from snow-capped peaks. While cotton clothing is lighter, woolens will keep you drier and warmer when the weather turns chilly. The layer system—with wool or wool-blend shirt, sweater and spinnaker shirt or windbreaker—is easier for a backpacker to put on or peel off as required rather than carrying a big expedition-weight parka. A head covering can help prevent loss of body heat, and a foldable hat or easily stowed knit watch cap is worth packing.

For wet weather, a lightweight poncho or plastic sheet should be part of your standard equipment. For most areas in summer, a sleeping bag that is comfortable down to 25° F. is suitable when used with a sleeping pad. Choose medium-weight hiking shoes that are well broken in and comfortable.

Meals should be planned around basic, easily prepared, dehydrated high-energy camp foods. Canned and bottled foods are usually too heavy to pack in comfortably. Repackage boxed sugar, cereal and dry milk in plastic bags. Water can be a problem; you should carry water-purification tablets when you camp back-country.

Your backpack seems to add weight the farther you hike and it is important to eliminate nonessentials. Don't carry a rifle unless you are expressly out for game during the hunting season. Instead of an ax, carry a good pocketknife or belt knife and a lightweight saw. Firewood is scarce in many wilderness and primitive areas; of course, only dead wood should be used for warming and cooking fires. Small pack stoves are a much better choice.

District rangers estimate that an average stay in a wilderness like Utah's High Uintas Primitive Area is about five days, and note that the amount of litter that can build up during those periods is enormous. The Forest Service encourages a "Pack it in—pack it out plus five (extra pieces of trash)" policy. All unused foodstuffs should be burned along with their wrappers and tinfoil. What doesn't burn to a crisp should be carried back out.

There are almost no toilet facilities in the back

**HIKING** through undergrowth of the 12 miles where the Linville Gorge encloses North Carolina's Linville River in the Pisgah National Forest, a camper finds no facilities for overnighting, but there are spectacular crags and rock formations, as well as dense foliage and 2000-foot elevations.

country, so spots well away from lakes, streams and trails should be used. Dig a shallow six-inch hole in the top layer of soil and then cover the hole with sod. Do not use soap or detergents near streams and lakes. Use a container to carry water for washing or shaving.

You will want to carry a camera and film, but choose a small light pocket-size model unless you are a professional photographer. Small flashlight, matches in waterproof container and first-aid kit also rate as essentials. Insect repellent may not be necessary for the area you choose, but it's better not to take a chance. Pack along some of the brand that works best for you.

Every area differs somewhat, so it is best to inquire ahead of time about a camping region that interests you. Here are a number of examples:

Information about California areas is available from:

**California Regional Headquarters**
**USDA Forest Service**
**630 Sansome St.,**
**San Francisco, CA 94111**

Among the 21 areas in the state are:

**Agua Tibia.** For the camper interested in flora and fauna, there is variety that ranges from desert plants at 1500 feet, a chaparral belt, and firs, cedars and pine up to 5000-foot level. Mountain lions, bobcats, deer, coyotes and many types of birds may be seen. Since much of the area has not had forest fires in over 100 years, many kinds of plants flourish. Because of fire hazard, use is limited to winter and spring; the area is closed from July to October.

**Marble Mountain.** Named for a mountain of white marble that gives the impression of perpetual snow, the area is forested and easy to hike over gently rising trails. Lakes are stocked with trout and the lower streams support ocean runs of steelhead and salmon. Wildlife is plentiful and deer and bear are hunted. Some rare alpine flowers are found only in this area.

**John Muir.** This 503,478-acre wilderness, largest in California, carries the name of the famed naturalist and extends along the crest of the Sierra Nevadas from Mammoth Lakes down to the Mount Whitney region. With thousands of well-stocked lakes and streams, this summer range for 50,000 mule deer is considered one of the most inviting wilderness areas in America.

For 15 areas in Washington and Oregon, address requests for information to:

**Pacific Northwest Regional HQ**
**Box 623, USDA Forest Service**
**319 S.W. Pine St.**
**Portland, OR 97208**

**UNRIVALED FISHING** in June Lake, with Carson Peak in the background, is one of the many attractions of the Minarets Wilderness Area in the high country east of Sierra crest in the Inyo and Sierra National Forests of California. Altitudes range from 7000 to 14,000 feet; temperatures from 85° to −15° F.

### WILDERNESS IN THE HIGH UINTAS PRIMITIVE AREA

Considered an ideal example of what a wilderness can be—remote, beautiful, unspoiled, challenging and little-known—the High Uintas Primitive Area in northeast Utah stretches for 243,957 acres up over the backbone of the Uintas Mountains and down into the Wasatch and Ashley National Forests. This mountain range, gouged and carved by at least three ice ages, runs east-west instead of the usual north-south.

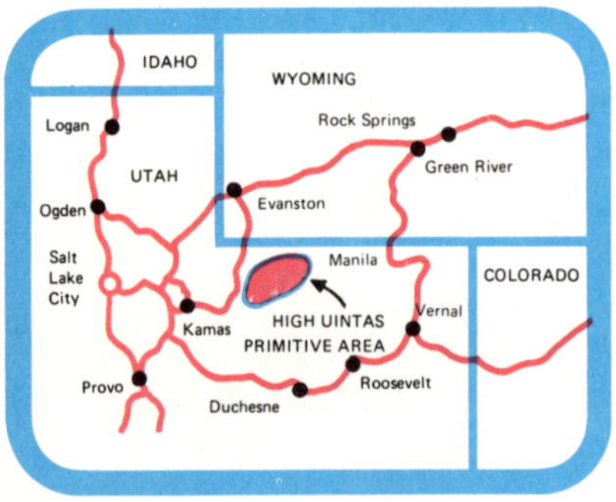

The slow-moving, grinding ice masses left behind thousands of high glacial lakes—probably only NASA Landsat satellite tapes show just how many. Campers are dwarfed by huge rock crags and spires, natural stone amphitheaters, ridges, canyons, mesas and icy streams that are born above 13,000 feet on high snow-covered peaks.

The Forest Service has a number of entry points where you can check in with a district ranger for maps and latest weather and terrain reports. An easy entry point is Kamas, 60 miles east of Salt Lake City. Mile-long Mirror Lake and surrounding campsites, meadows and streams make it popular, and there are 78 smaller lakes within a radius of six miles. Though considered "family country," there have been mountain-climbing fatalities in the vicinity, and the first day can be well spent lounging, plotting trails on area maps and becoming acclimated to the 10,000-foot altitude.

District ranger Steve Scott suggests newcomers pack along sections of the well-marked Highline Trail which winds 56 miles through the mountains to the Henry's Fork entry point in Wyoming. Plan to cover no more than two miles an hour when packing at these altitudes; temperatures range from a high of 70° F. down to 30° at night and you'll need snowshoes until mid-July. For hiking, figure you'll burn 5000 to 6000 calories a day and should pack in a dry weight of at least 1½ pounds of high-energy food for each day of camping. A light pack rod can land brook, rainbow and cutthroat trout or arctic grayling.

While rangers do all they can, there are places in the HUPA where a camper in trouble can be taken out only by helicopter—a chancy and expensive way to end a holiday. Careful preplanning is important. For High Uintas information, contact the ranger district at Duchesne, Kamas, Dutch John, Roosevelt or Vernal, UT; or Evanston or Mountain View, WY.

**Mount Hood.** Among snow-capped mountains, Mount Hood may rank next to Fujiyama for the greatest number of climbers. Novices with guides and rented equipment can reach the summit in the summer, but the weather is unpredictable and climbers should sign in and out at registers at three ranger stations. Trails are usually opened for foot travel around July 15, although they may not be free from snow until mid-August. Wildlife is not plentiful here, but deer, bear, bobcat, squirrels, grouse and cougar may be seen.

**Paysayten.** Almost every wilderness experience can be found in this 20 by 40-mile region of the Mount Baker and Okanogan National Forest. The wide range of topography, plant cover and elevations offers a guided horseback tour for a senior citizen or a challenging climb for a mountaineer. Most of the 94 mountain lakes are stocked with cutthroat, eastern brook, Montana blackspot, rainbow, German brown or Dolly Varden trout. There is seasonal hunting for deer, mountain goat and bear.

Inquiries about the Colorado, Idaho, Montana, Nevada, Utah and Wyoming areas can go to:

**Rocky Mountain Regional HQ**
**USDA Forest Service**
**Federal Center Building 85**
**Denver, CO 80225**

**Maroon Bells-Snowmass.** Picture postcard peaks fill the skies of this popular region that is easy to reach from boundary-line campgrounds in the White River National Forest. Mountains and high ridges range from 12,000 to 14,259 feet. Even hot springs gush out near timberline where mountain sheep summer and winter.

Information about Montana and Idaho areas is available from:

**Northern Regional HQ**
**USDA Forest Service**
**Federal Building**
**Missoula, MT 59801**

**Bob Marshall.** Extending 60 miles along the Continental Divide, this rugged back country has game that include wolverine, grizzlies, deer, elk, moose, black bear, mountain goat, mountain sheep and mountain lion. Fossils from animals of a billion years ago are also found here.

Arizona and New Mexico, with 17 areas, offer camping wilderness managed by:

**Southwestern Regional HQ**
**USDA Forest Service**
**517 Gold Ave. S.W.**
**Albuquerque, NM 87101**

**Gila.** This area contains the first public lands ever set aside because of their wild unspoiled quality. The desert canyons here were once the hideout of an earlier camper, Chief Geronimo.

Most wilderness and primitive areas are in the west, but within the last couple years a number of additions have been made in the north and southeast. Information is available from:

**Eastern Regional HQ**
**USDA Forest Service**
**633 West Wisconsin Ave.**
**Milwaukee, WI 53203**

**Boundary Water Canoe Area.** This famous region is very different from any other in the National Forest Wilderness System. Its 747,840 acres along the Minnesota-Canada border are one-fifth water. Over 1000 lakes of 10 acres or more in size are waiting here, plus a lacework of rivers and streams. Here jagged rocks and lush undergrowth cover islands edging waterways once paddled by the Chippewa and Sioux, early voyagers who trapped and traded furs. Today's camper can fit out with canoe and gear at a town like Ely and then paddle in away from it all. Unlike other wilderness areas, limited timbering also goes on in portions of the region.

The 1950s and '60s saw land set aside to be forever wild in North Carolina and New Hampshire.

**Linville Gorge.** Steep slopes and overhanging cliffs enclose 12 miles of the Linville River as it drops 2000 feet into valleys below. Word should be left with a district ranger before you enter to camp this rugged territory. Shining Rock, with elevations from 3500 to 6033 feet, and outcroppings of white quartz, forms another wilderness of 13,350 acres within the Pisgah National Forest in western North Carolina.

**Great Gulf.** In New Hampshire's White Mountain National Forest, on the eastern slopes of the Presidential Mountain Range between Mount Washington and the Northern Peaks, a 1600-foot-deep valley extends east for 3½ miles. Gouged out by ancient glaciers and ravaged by hurricanes that have torn the timber of the upper slopes, this 5552-acre area offers the hiking camper a rugged climb through growth that ranges from lush up to barren. Two lakes are in the upper reaches of the area. The region seems like a small oasis in the heavily populated East.

Added to the two areas in North Carolina are a dozen new regions for which camping and hiking data can be obtained from:

**Southern Regional HQ**
**USDA Forest Service**
**1720 Peachtree Rd.**
**Atlanta, GA 30309**

These areas are now open: the Upper Buffalo and Caney Creek Wilderness Areas in Arkansas; Sipsey, AL; Bradwell Bay, FL; Ellicott Rock in North and South Carolina and Georgia; Cohutta in Georgia and Tennessee; Gee Creek, TN; Joyce Kilmer–Slickrock in North Carolina and Tennessee; Beaver Creek, KY; James River Face, VA; and Dolly Sods and Otter Creek, WV.

1 All the equipment above fits neatly aboard and out of sight in special shipshape shelves and cabinets.

2 Photography can be a moneymaking hobby. Complete darkroom equipment fits safely under one dinette seat.

3 Top of an old dresser is a small table for typing travel stories; it stays permanently in place.

4 Envy of other campers is galley toolboard that holds short-order utensils in place with shock cords.

# A camping expert shows how to fit out an RV

By E. P. HADDON

■ PLAN AHEAD, and you can fit out your recreational vehicle for almost any activity—or a dozen of them. Take advantage of unused space, and you'll be able to stow the essential gear for boating, fishing, hunting, photography and hiking. The assortment of gear that fits in my economy-size rig continues to amaze owners of highway dreadnoughts twice its size.

Shipshape and secure storage is the secret. The first shakedown cruise with our latest rig showed that our 11 cabinet doors offered easy access to the overhead compartments for stowing groceries and other small items. After a day on the move, however, nothing remained where it had been placed.

The first remedy was to install ¼-inch-plywood partitions to divide the spaces inside each compartment and assure a permanent berth for everything. Dishes are now safely held in cushioned racks designed for RVs and screwed in place. Glasses ride on other racks attached to the inner sides of the cabinet doors.

Opposite the dinette and beside the combination bunk/seat, an 18 x 36-inch oak table (once a discarded dresser top) fits neatly atop a single folding leg. It's an ideal accessory worktable, does double duty serving guests in for dinner or a drink, and has become so indispensable that we leave it permanently in place. On the underside of the table, a shallow partitioned tray provides a handy hiding place for innumerable small objects. Behind, on the wall, are two magazine racks.

Items originally stowed inside the two dinette

SEE ALSO
**Backpacking . . . Bicycles . . . Binoculars . . .
Boat camping . . . Boating, white-water . . .
Campers . . . Canoes . . . Emergencies . . .
Exercising . . . Hunting . . . Map reading . . .
Recreational vehicles . . . Trailers, tent . . .
Vacation homes**

seats were invariably lost in a jumbled hodgepodge. Two new plywood partitions create three compartments in one.

Beneath the other dinette seat is a complete hobby darkroom kit. The rather delicate enlarger head is cradled in plastic foam. One corner is reserved for photographic chemicals, all carried in unbreakable plastic bottles.

For outdoor eating and activities, we carry a small 15 x 33-inch unit that unfolds from 4 inches thick into a table with four attached seats.

A surprising amount of usable space was retrieved by fitting a simple 44 x 28 x 10-inch cabinet into wasted space below the refrigerator and clothes closet. Provided with shelves and sliding doors, it's just right for storage.

To make RV cooking easier, all short-order essentials are stored on an enameled perforated board on the wall behind the sink and stove area. Utensils and accessories are held in place while we're on the move with Coghlan's Mini Stretch elastic shock cords, available at camping supply outlets.

Since we didn't need the extra-long bed with which this cab-over mini-motor home came equipped, I used an electric carving knife to make quick work of slicing 14 inches from one end of its foam mattress. Now we still have a full-size double bed, plus room for yet another invaluable cabinet.

Neatly strapped into place on the inner side of its doors are a .22-cal. Colt Matchmaster target and hunting pistol, flanked by Buck hunting knives, a hatchet and other accessories.

Enlarged exhibition-type nature photos that I have taken are mounted on the cabinet doors and have been admired by visitors.

Before I revamped the outside storage cabinet, equipment inside always seemed difficult to find. Again, I installed a shelf and one dividing partition, and now everything seems easier to reach.

Though the rig's roof deck permits toting the 17-foot folding kayak topside, I usually carry it inside the vehicle. A slim lightweight bag containing the kayak frame rides on the floor beside the bunk seat at the rear. A bag with the folded hull stands inside the door without blocking the entrance. An inflatable kayak lashed to the rear of the right front seat is out of the way.

Ingenuity rather than cost often determines the ways you can fit out your motor home or trailer for adventure.

**Overhead cabinets with added partitions and shock-cord tie-downs stow gear securely.**

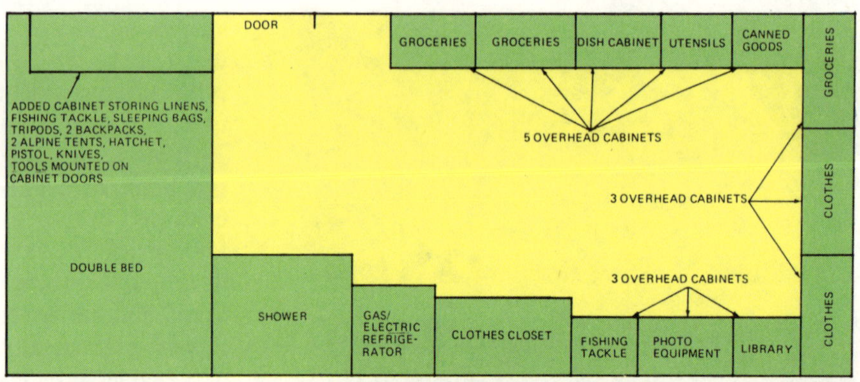


**Undercounter bins plus compartments below dinette and bunk seats hold supplies.**

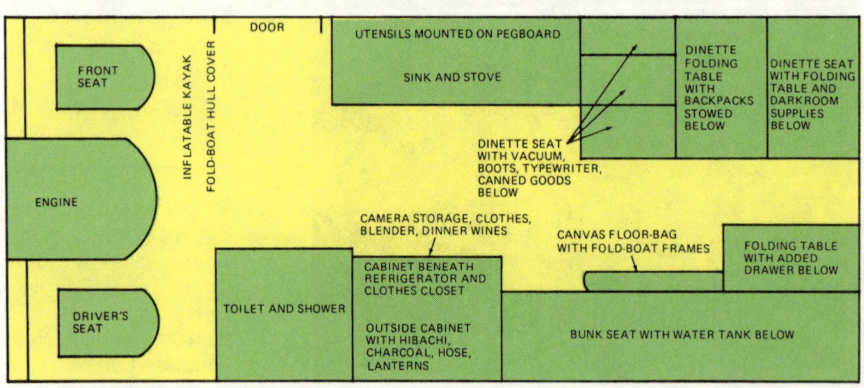

**WHEN LAMINATING** two or more plywood panels, here's a suggestion for providing clamping pressure at the center. According to the folks at 3M Company, common C-clamps will do the job for you. After spreading glue on mating surfaces, position the panels and run a cleat down each edge. You then position a third, thicker cleat down the center. Use scrap wood to span the cleats.—*Bob Brightman*.

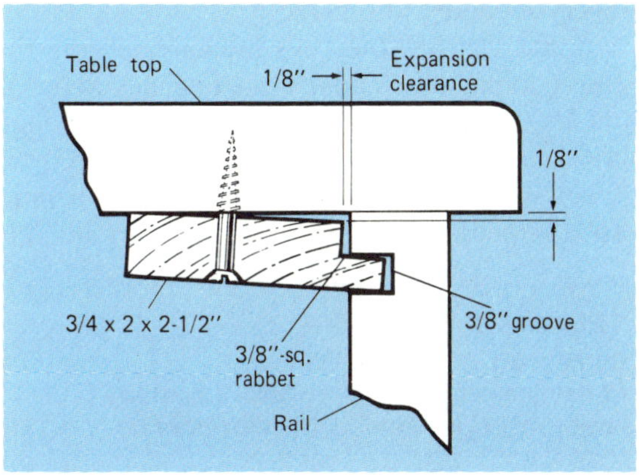


**A METHOD OF FASTENING TABLETOPS** which I've used for years with great success is illustrated in the drawing to the left. It's important that you allow sufficient clearance between the table and the rail. Also, when installing the fastener, leave about ⅛-in. clearance so that the screw will pull the parts together snugly.—*Harold R. Crawford*.

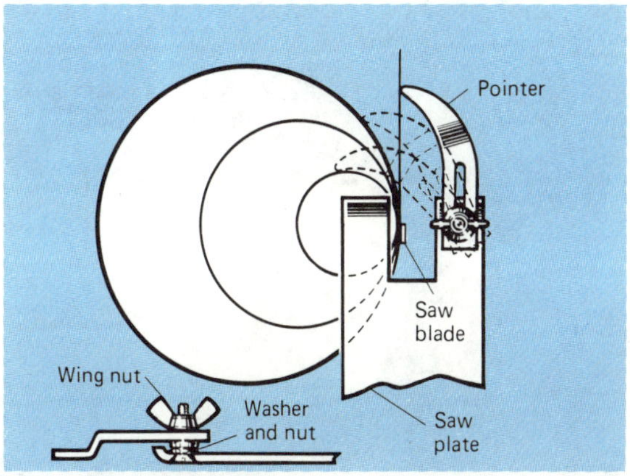


**WHEN CUTTING** with a sabre saw, most persons find it hard to follow a line freehand. To make this task easier, I added an adjustable pointer to the saw shoe as shown in the illustration at the left. Simply start the saw on the line you want to cut, turn off the saw and set the pointer on the line. The attachment will let you make a much neater cut.—*H.C. Anderson*.

**AN ELECTRICAL OUTLET** can easily be created where you need it in your workshop. You simply need to splice into a BX cable strung below a joist and install an outlet box. After you cut off the power, cut the cable and install the box on the joist. You'll find there is plenty of slack in the line to permit the installation. Complete the job by adding the plate cover.—*Bob Brightman*.

# Get started in candlemaking

By WAYNE C. LECKEY

FANCY CANDLES make perfect Christmas gifts—they're colorful, decorative and useful; they symbolize Christmas, and they're gifts that everyone likes to receive.

What's more, they are homemade gifts that don't look homemade. Cast in fancy reusable molds, they have a ready-made, store-bought look that compares with candles which bring fancy prices in gift shops. Yet they cost pennies to make and anyone can make them. The

SEE ALSO
**Bottle craft . . . Candle stands . . .
Christmas decorations . . . Gifts, Christmas . . .
Jewelry . . . Lapidary . . . Stained glass**

**MELT WAX** in a double boiler, never over direct heat. Melted wax should be no hotter than 175-185°F.

**ADD COLORING** after wax is completely melted. Stir and allow the coloring to dissolve thoroughly.

"tools" required are common household items for the most part, and hobby stores provide a ready source for the molds, wax, wicking, scent, coloring and other items. And once you're hooked by this fascinating hobby, you'll soon be making your own molds from Ultra Cal and Silastic RTV compounds. You'll be creating such unique forms as ice candles, whipped-wax candles, sand candles and others. You'll be decorating them in gold, spray painting them in pearl and iridescent tones and adding glitter to make them sparkle. There are many good books on candlemaking which take you beyond the basic steps explained here.

**kinds of molds**

Candles can be cast in countless sizes and shapes, and in several kinds of molds. Flexible, rubberlike stretch molds are one-piece molds which you remove from the finished candle by pulling them off like rubber gloves—inside out. Such molds, like others, can be used over and over.

Vacuum-formed plastic molds are two-part molds of matching halves that are self-aligning but must be sealed at the edges with masking tape.

Metal candle molds, straight and tapered, are available up to 18 in. tall. Handsome one, two and three-wick candles, round, triangular and star-shape, can be cast in metal molds.

**wax, wick, coloring and scent**

Candle wax comes in slab and granular form, and can be melted and remelted again and again.

Wicking comes in spools of 100 and 300 ft.,

**TWO-PIECE** plastic molds consist of matching halves bound and sealed at the edges with masking tape.

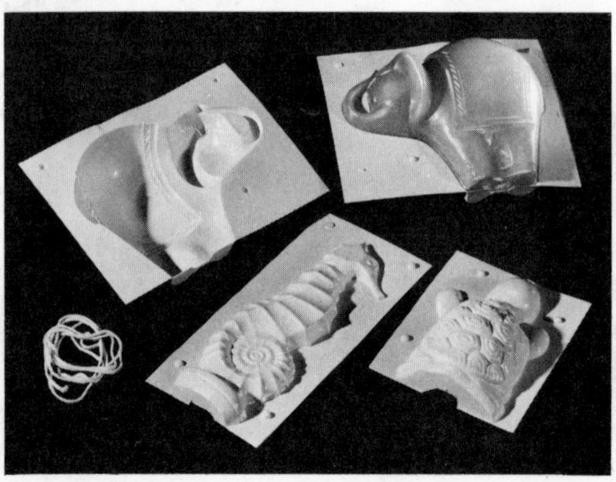

**ONE-PIECE** stretch molds are flexible. To remove from the candle, you pull them off like a rubber glove.

A SMALL VISE and spring clamps are handy for holding together the halves of a large plastic mold. A notched brace helps hold it steady for pouring.

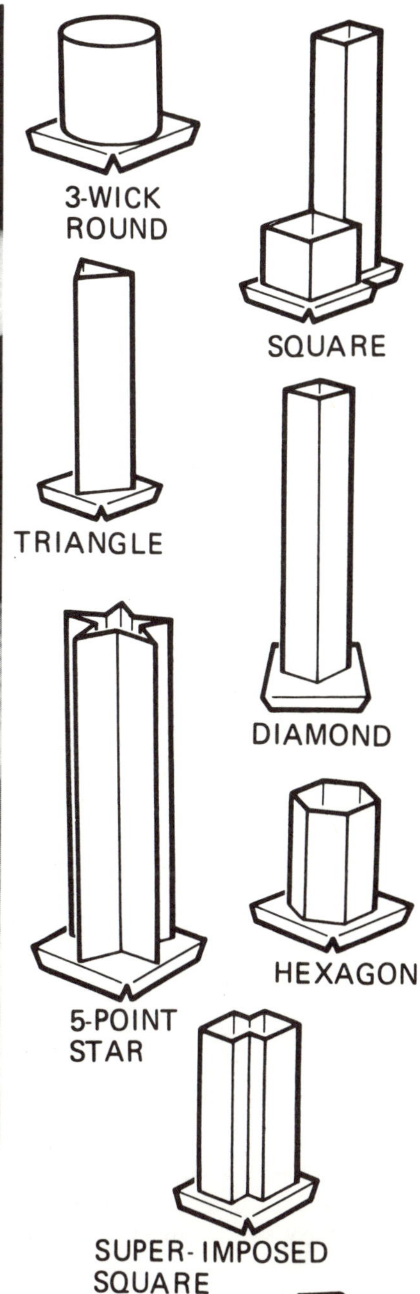


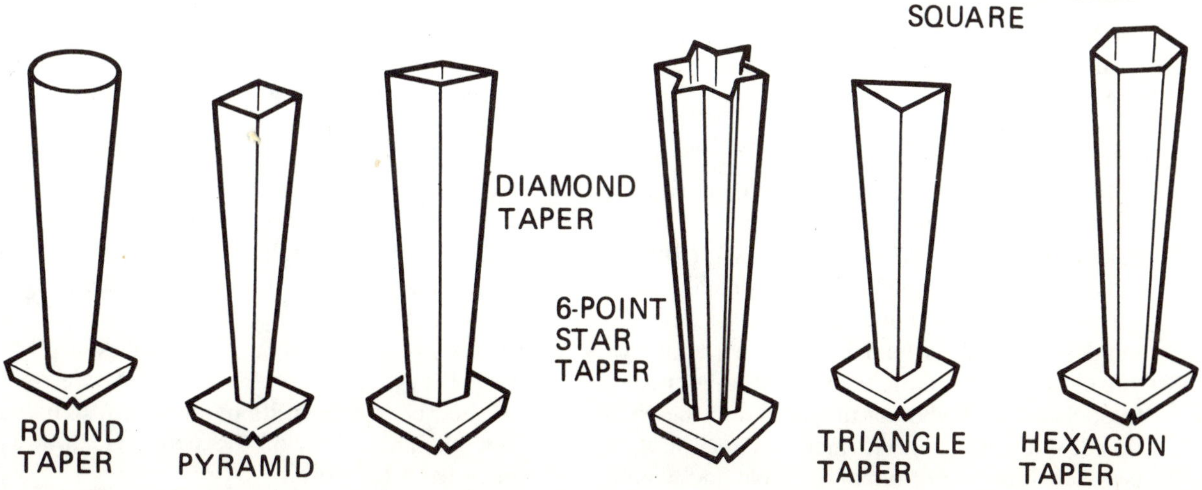

also in 6 and 12-ft. lengths, braided or cored. Braided wicking is specially made for use with metal molds, whereas core wicking is for non-metal molds.

Coloring is sold in square, granular, powder and stick form. It's added to the melted wax before pouring and allowed to dissolve completely. It comes in colors of the rainbow, and color is controlled by the amount that's added to the wax. Regular coloring-book crayons also can be used to color candle wax.

Similarly you can add fragrance to your candles. Available in some 17 different scents, it's added to the melted wax after the color and just before pouring. A ¼-oz. bottle will scent 5 lbs. of wax.

**melting the wax**

Candle wax is melted in a double boiler, never over direct heat. A fruit-juice can, or a 1-lb. coffee can, fitted with a suitable handle and pinched at the rim to form a pouring lip,makes a good ladle in which to melt wax. It's best to have a couple of such ladles at hand so you can have two double boilers going at one time. Bring the water in your outer vessel to a slow boil, then turn the heat down to a medium simmer. Wax should be about 175-185°F., no hotter. You can test it with a candy thermometer.

**setting up your mold**

One of the best ways to support a rubber stretch mold is to rest it in an inverted cardboard paint-bucket liner. A hole is cut in the bottom to accommodate the mold, and the mold is suspended in the hole. In the case of two-piece plastic molds, the two halves are sealed around the edges with masking tape (after the wick is in place, of course) and the mold is supported vertically. A simple way to do this is to use a small vise which clamps over the edge of a table, and secure the mold in it.

Metal molds are no problem: they're free-standing and can't tip over.

**inserting the wick**

In most cases, depending upon the particular mold, the wick is placed in the center and supported at the top. When working with rubber molds, the wick can be held with a pencil placed across the top. In the case of two-part plastic molds, the wick is held in place with tabs of masking tape. Here the wick is cut long enough to stick out both ends of the mold and the ends taped. When working with metal molds, the wick is attached to the bottom of the mold and wound around a rod bridging the top. In each case the wick is first dipped in melted wax and allowed to cool before it is centered in the mold. Predipping the wick will make it burn better.

Melted wax is hot so take care in pouring it into your mold. Also, avoid getting any water in the wax since it will cause bubbles in the finished candle. Fill the mold almost to the top and let stand. As it cools, wax shrinks, so when it starts to get a crust, poke a hole through the hardened wax and add more melted wax to bring it level. You may have to do this several times as the wax continues to shrink and form air pockets. It requires about three hours for the wax to harden. To cure, place the mold overnight in a refrigerator. Chilling will also give added shine to the wax.

When you are able to tilt the mold while filling it, as you can with metal molds, doing so and gradually bringing the mold upright as you pour helps prevent the wax from becoming cloudy due to excessive air bubbles.

**removing candle from mold**

A flexible mold is removed from the candle much the same way as pulling a rubber glove from the hand; it's actually turned inside out. Dusting the outside of the mold with talcum will keep it from sticking to itself.

To open a two-part mold the tape is peeled from the edges and the halves gently pulled free. If they stick, hold the mold under hot water for a few seconds. Spraying the mold beforehand with silicone or coating with cooking oil will ensure easy release. To remove a candle from a metal mold you simply turn the mold upside down and tap the bottom lightly. A knife can be used to pare and smooth any unwanted ridges.

Finishing off the bottom of candles so they sit level can be done in two ways. You can shave off the wax with a rasplike tool called Surform, or you can rub the candle over the bottom of a pan placed over low heat on your rangetop.

Novelty candles in the form of frogs, turtles, owls, seahorses, snowmen, mushrooms and the like are especially fun to make, as are candles that float. New for a garden pool or romantic table centerpiece, lighted floating candles are fascinating to watch with their ever-changing reflections. Pictured on page 559, the candles resemble lily pads. They are cast in flexible molds while they rest in pans of water. You release the finished candle from its mold by gently *pulling* (not pushing) each of the petal "separators" *away* from the core.

**NOVELTY CANDLES** make popular gifts for Christmas, Easter and other holidays.

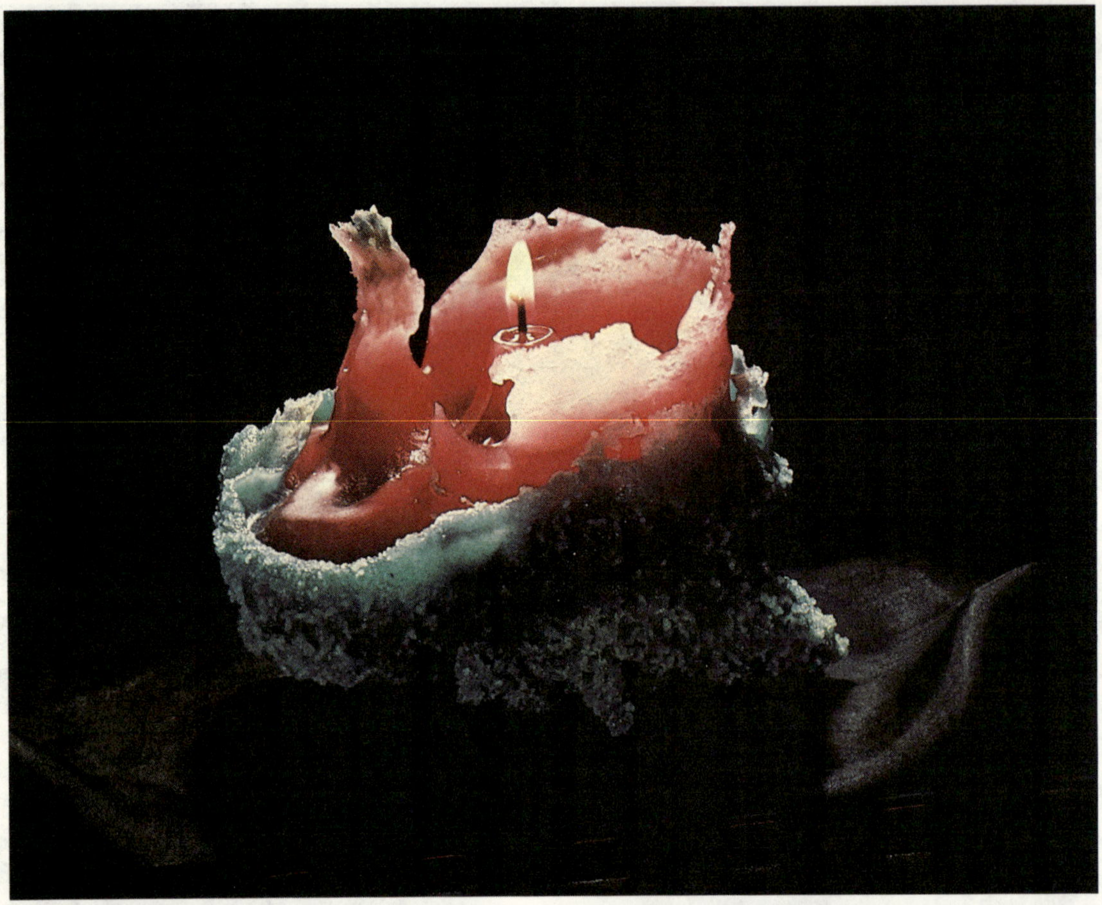

**DELICATE, FROSTY LOOK** is special feature of snow-molded candles. The one at the top of the page was made with two pours of different wax colors. Each of the three candles above was made with a single pour of wax. Two of the candles above are "upside down" candles. Another two-pour candle, right, shows the delicacy—and unpredictability—of the results. No two snow candles can ever be exactly alike.

**UNIQUE CANDLES** are created by pouring wax into a snow "mold." Resulting designs are as varied as snowflakes.

# Free-form candles molded in snow

**Just as no two snowflakes are ever alike, no two candles molded in snow are alike. Here's how to make these delicate candles in one or multiple colors**

**By DOUG DILL**

SEE ALSO

**Art glass . . . Bottle craft . . . Christmas decorations . . . Coasters . . . Gifts, Christmas . . . Jewelry . . . Lapidary . . . Marquetry . . . Plaster casting . . . Velvetizing**

■ SNOW CAN BE the mold you use to create candles that take on a frosty appearance, yet provide a warm glow for winter nights. Hot wax melts and changes its snow molds as it solidifies, producing unusual textures and shapes. And with snow molds, no two candles are ever alike.

Snow candles are made with your favorite brands of candle wax, colors, scents and wicks. Snow becomes the mold. The process is like making sand candles except that the snow molds are not as predictable as sand and can result in pleasant surprises for the candlemaker.

If you live in snow country you can melt the wax in the normal fashion on your home stove. If you have to travel to the snow, a campstove makes a fine source of heat. In either case, use a double boiler, never direct heat.

Start making your molds in the snow while the wax is melting. Simply create a depression that wax can be poured into. Dangle a wick into the depression from a stick placed across its rim; the wick should be coated with melted wax and allowed to dry thoroughly before being placed in the mold. Wire-core wicks seem to work best.

There are two basic types of snow candles. The "upside down" candle is made by sculpting a design in the bottom of your depression in the snow. After the wax is poured and set, the candle is removed and turned upside down for use.

More unpredictable designs happen with the two-pour candle. After one pour has solidified completely, more wax, usually of a second color, is poured into the mold. Results can be surprising. Instead of filling the mold to the top, the second pour will run to the bottom of the mold and harden under the first, making its own design. This candle is not inverted, but used in the position in which it is formed. Usually some wax must be removed from the bottom to make it sit flat; use a hot knife and be careful not to break off delicate parts. Sometimes the second pour will go deeper than the wick and another must be inserted; use a heated icepick to make a hole, place the new wick and pour melted wax around it.

It takes about an hour for wax to harden enough for removal from the snow. Be very careful when digging out the candle—parts may spread far from the original mold and can easily be broken.

Snow characteristics will affect candles. Try molds made in new or soft snow, both packed and unpacked. Crusty or melting snow makes interesting but generally more predictably shaped candles.

# Holiday candle stand

**This decorative stand holds jumbo candles as a Christmas centerpiece. It is made from ½-inch stock and all the legs can easily be cut at once with a band saw**

By ELMA and WILLARD WALTNER

DESIGNED TO HOLD jumbo candles up to 3 in. in diameter, this attractive candle stand makes a festive centerpiece for the Christmas holidays.

First make a full-size pattern for one of the arms and trace it on your wood. If you have a bandsaw, you can saw all six arms at one time by tacking the ½-in. stock together in the waste areas. A router with a corner-rounding bit makes quick work of rounding the edges of the arms, but you also can do it with a contour sander or a sanding drum in the drill press. If you have to do it by hand, start with a coarse sandpaper and finish with fine.

The candle cap is lathe-turned from a ¾-in.-thick disc and recessed ⅛ in. Again, if you don't have a lathe, it can be built up from three jig-sawed discs.

If you have used open-grain wood such as mahogany or walnut, apply paste wood filler before assembling the parts. Glue and nails are used to attach the arms to the cup. Add two arms at a time, one opposite the other, and clamp. Complete the stand with self-rubbing clear satin finish.

SEE ALSO

**Bandsaws . . . Birds, carving . . . Centerpieces . . . Christmas decorations . . . Gifts, Christmas . . . Wood sculpture**

**AFTER SAWING OUT** the arms, all edges are rounded with a Sand-O-Flex contour sander or sanding block.

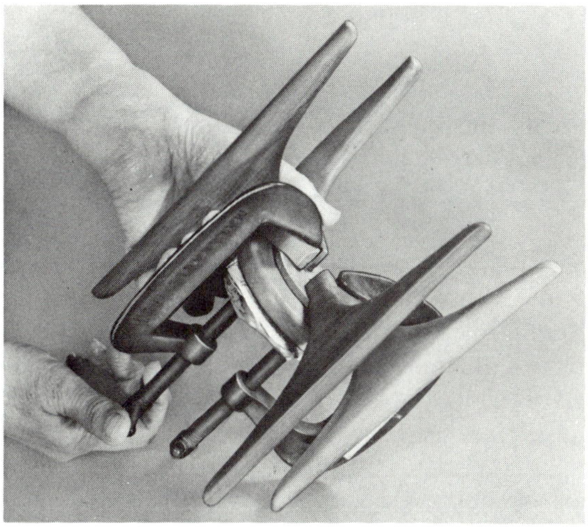

**THEN THE ARMS** are glued and nailed to the bottom of the candle cup and clamped until the glue dries.

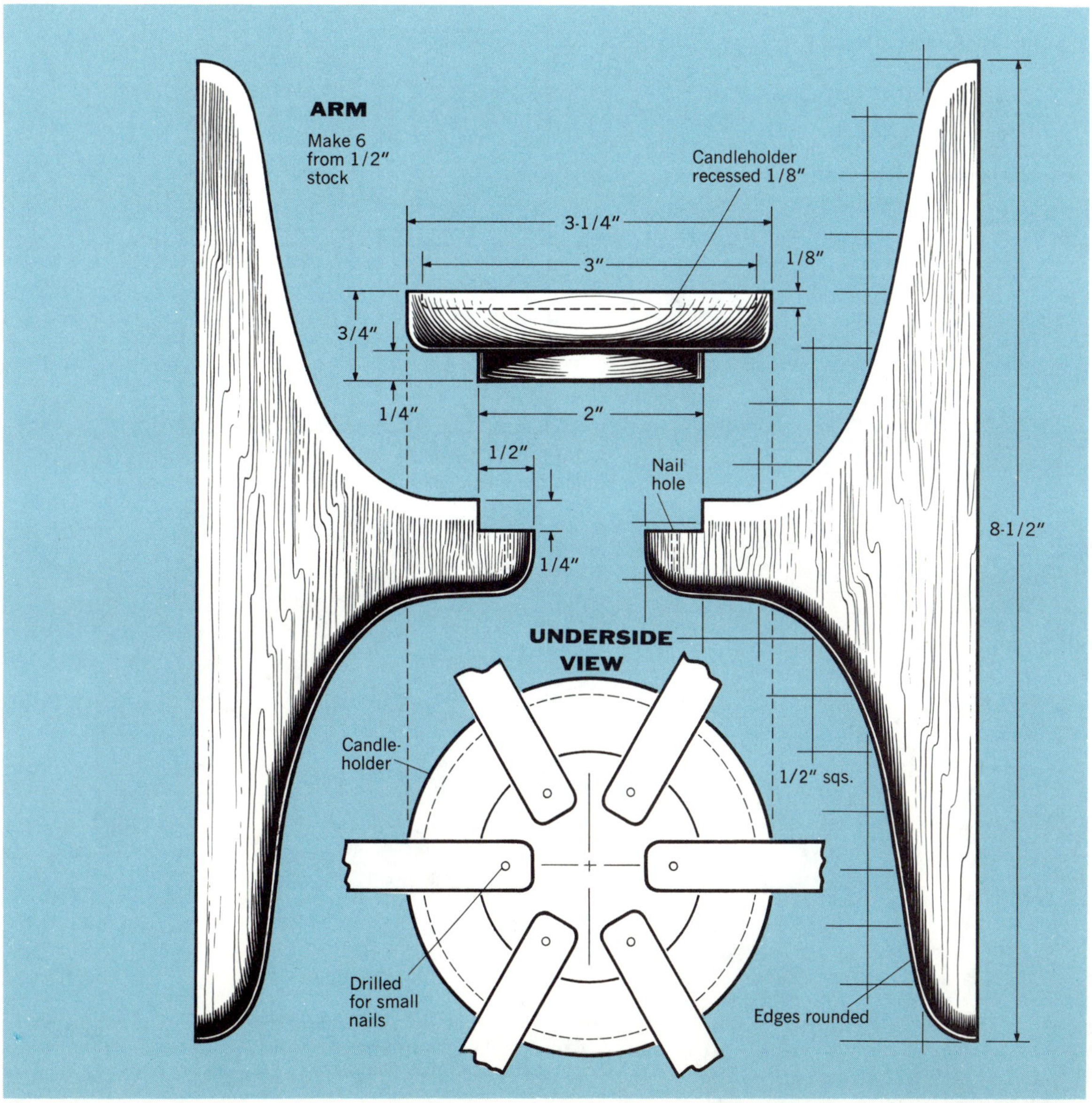

# Flamed candle stand

By WILLARD and ELMA WALTNER

■ THIS UNIQUE candle stand begins as a 3½x3½x4-in. redwood cube. Turn the wells for the three sizes of candles in the top plane of the cube, following the dimensions given. Trace the carving designs on the four sides of the cube so that like designs are on sides opposite each other.

Before flaming the block, recess the background ⅛ in. deep with carving burrs in a drill press. Clamp a piece of wood to the drill-press table to make the straight outer cuts, and clean out the rest of the wood freehand. With the cube clamped in a vise, use a propane torch at low flame and pass it back and forth over the work so the tip just touches the wood to char it lightly.

If the wood should catch fire, blow it out. If necessary, remove heavy charring from the cube with a wire brush. Move the brush with the grain, then rub the block with a soft cloth and apply two coats of a satin-finish clear lacquer.

**BLOCK IS MOUNTED** in four-jaw lathe chuck and wells are turned in top for three sizes of candles.

**BACKGROUND OF DESIGN** is carved freehand to a depth of ⅛ in. with a burr cutter in the drill press.

CARVED PATTERNS

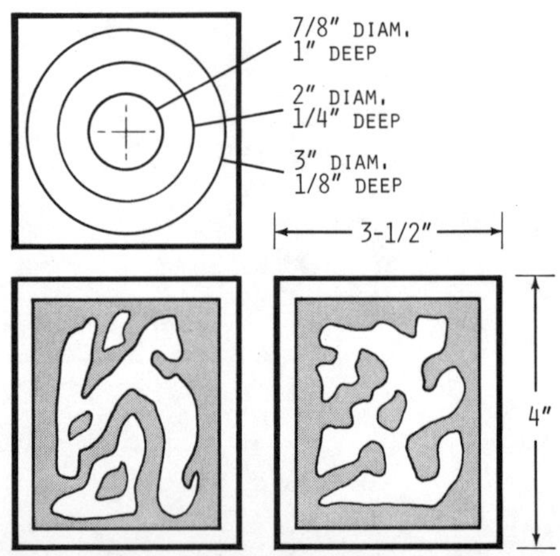


**BOARD IS CLAMPED** to a drill-press table for a fence to guide the block when you make straight burr cuts.

SEE ALSO

**Centerpieces . . . Drill-press techniques . . . Gifts, Christmas . . . Internal carving . . . Tools, hand . . . Wood sculpture**

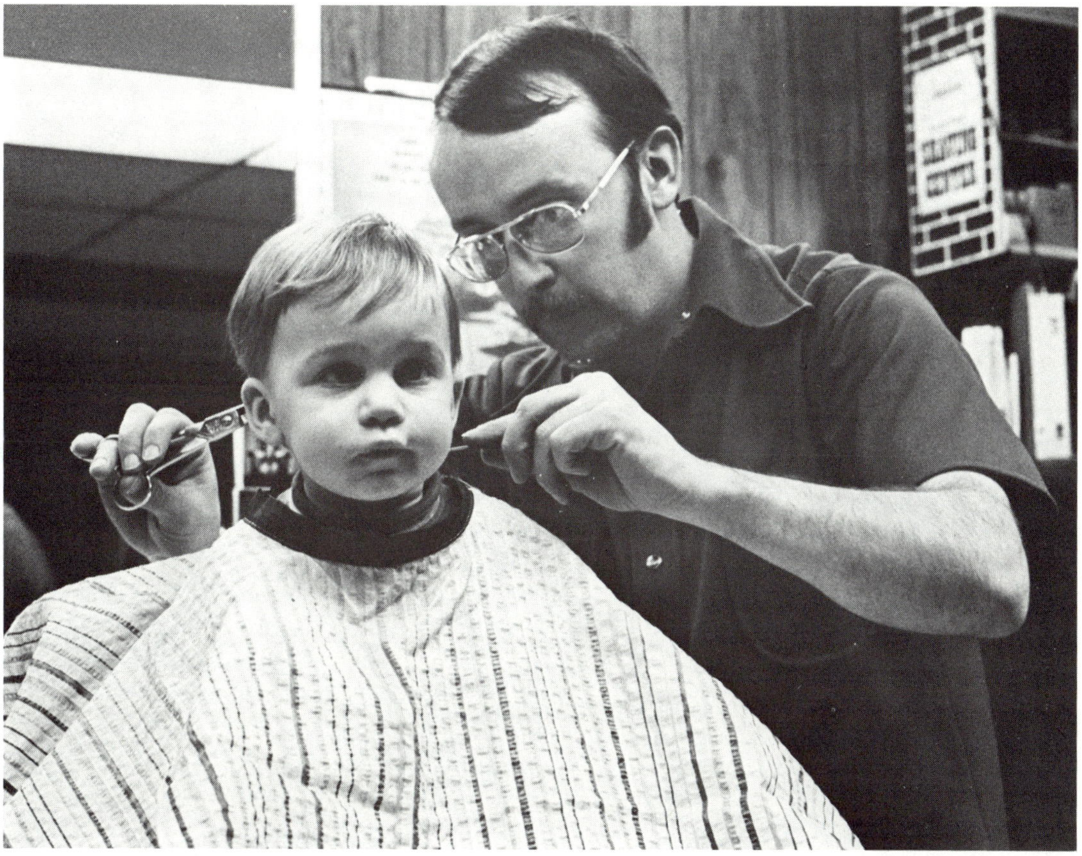

**NEIGHBORHOOD ACTIVITIES** make good picture possibilities you'll enjoy in later years—especially if your own family is participating. But click off enough frames to let people get used to your shooting

**DOING THEIR** own thing in their own homes, your subjects feel more at ease—and the home environment supports the picture. We still have the afghan Grandma made here

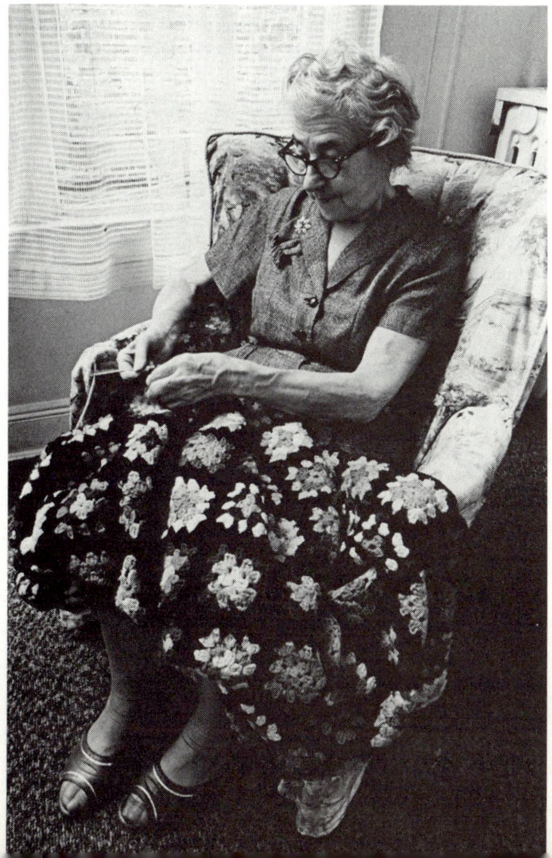

# Family photos can be more than snapshots

By DAVID SAGARIN

DON'T SNEER at the snapshot. It's a quick response—aim and shoot. But when everything works right, there's a wonderful freshness and reality to it. If you're not quite ready, though, or something isn't working, you're out of luck, and the moment passes.

A formal photograph, with properly adjusted lighting, carefully determined exposure and traditional posing, generally comes out all right but without much individuality. Sometimes you feel

**STAND BACK** when shooting people in activities—but just far enough to make the activity and the people's relationship to their environment clear. There's not much doubt about where this activity is taking place

**GET IN** close and fill your frame with your subject. You don't even need a "portrait" focal-length lens: This shot was taken with a normal 55-mm lens at about 2½ feet from the subject. You should experiment with your camera to determine the bes distance for this kind of portrait

more comfortable giving traditional photos, but there's a quality to a really good candid shot you can't get any other way.

What's a candid? It's a *planned snapshot,* made by predicting the course of spontaneous action and preparing yourself to take the best advantage of it.

Your planning needn't be complex—it can't be. Just try to picture in your mind what the finished print or slide will look like: Relate the situation to other times you've shot in similar light, or with similar backgrounds. If there's something you weren't happy about then, change it now. Maybe direct sunlight caused a squint, or a background of tangled foliage was confusing. Or maybe every time you're at the beach you overexpose (or, with metered cameras, underexpose).

You can control these things by moving around, by getting higher or lower, or by fudging the exposure settings in the direction you think they should go. You also gain control from prediction: Move to get a good background—*not* for the action as it's happening now, but for the action where you judge it will be happening by the time you've finished moving.

And instead of trying to keep action in focus as it moves, prefocus where you know the action's going, and shoot just as it gets there.

When the kids are doing something great, don't worry that if you take the time to change position you'll lose your shot. Shoot quickly where you are, then shift and shoot again—film is cheap, but memories are priceless. And once you start thinking of positions as you shoot, you'll find yourself already in a good position when the best action starts. Not only that, but as you get into what you're doing, you'll be less of a distraction to the people you're trying to shoot.

You might think it's easier to shoot good informal portraits of people you know well, but it's harder: The relationship gets in the way. A very fine professional photographer I know always took awful pictures of his wife—awkward, ugly, harassed. But that was his view of her; they're divorced now, and he takes lovely pictures of his new girl friend.

As long as you're aware of this problem, you can fight it. Just deal with those close to you as objectively as you can in the shooting situation. Concentrate on trying for good pictures, and they'll *be* good pictures.

**AN UNUSUAL** pose and activity kept my subject's interest alive, for a livelier expression. By getting above the subject I was able to get a simpler background than if I had shot her against the branches of the tree

**SHOOTING** kids in low-light situations or indoors? Get them involved in activities where slow shutter speeds and slow focusing won't be problems. And don't feel that you have to show a face in every shot

■ Art and technique are blood brothers, not enemies. The more confident you are in the technical aspects of photography, the more forethought you've lavished on them, the more attention you'll be able to give to the situation, people, composition . . . *pictures.* When you see a photo you really like, think about how it was made: Where was the photographer in relation to the subject? What lens was used—telephoto, normal, wide-angle? What would the effect have been if another lens or viewpoint had been used? What's the relationship between photographer and subject?

Look hard at lighting, too. Compare your memories of good pictures with the lighting you're about to shoot by: Your lighting can usually—and easily—be improved. In bright sunlight, the first consideration is to lighten the shadows and soften the modeling of features with reflected "fill" light. A light wall (not a bright color, unless you're shooting black-and-white) will serve; so will a white card or cloth. Better yet, work in open shade to minimize skin texture. With black-and-white film, increase the ASA rating about 50 percent and double the development time. This will increase contrast enough to offset the flat light and lighten skin tones (which also helps make blemishes disappear).

Indoors, you get more natural results with bounce flash or available light, but the same basic principles hold: Keep lighting contrasts reasonably low, and fill in shadows that contain important subject detail.

But indoors or outdoors, plan your snapshots as you shoot, and they won't be just snapshots—they'll be photos you'll be proud to show as photos, as well as captured memories.

Getting good candids often means shooting a lot of film—but it does not mean showing lots of pictures. Even within the family, too many shots don't preserve memories; they dissipate them. So once your shots have been developed, edit ruthlessly until you've come up with the one or two shots that best show the people and the situation (but don't be afraid to do a mini picture story of several related, but unduplicated shots).

You can edit parts of pictures, too, either cropping as you enlarge or trimming off unwanted details from store-bought prints. But better yet, think as you shoot—and edit in your viewfinder.

The caned seat looks expensive and difficult to weave, but it isn't. You can do one in a few hours for under $5

# How to cane a chair

By LEN HILTS

SEATS AND BACKS of woven cane have been used in fine furniture for several centuries. They look good, they're durable, and they're comfortable to sit on. But in time, the cane becomes worn and breaks, and when this happens there is little you can do to repair it. Your only option is to install new cane.

Because the pattern of the woven cane looks intricate, most people assume that the weaving is a job for experts only. Actually, cane weaving is a fairly simple operation, requiring no special tools, and no great skill or strength. You need only the materials, available from a dozen or more good mail-order sources, and the patience to follow a prescribed series of steps.

Cane is the outer bark of the rattan palm found in India, China, Ceylon and Malaysia. It is cut into strips varying in width from 1/16- to 3/16-in., and bundled into hanks of 1000 feet, with individual strips in the hank ranging up to 18 feet in length. You can also buy 500-foot hanks.

You pay a small premium for selected cane of higher quality, and an additional premium for select long cane, where every strip in the hank is more than 12 feet long. As this is written, a standard 1000-foot hank costs $7.50 to $8.00.

SEE ALSO
Armchairs . . . Chairs . . . Furniture, care of . . .
Legs, furniture . . . Sofa-beds . . . Stools . . .
Upholstery . . . Wood bending

Premium quality may run $10 a hank, and the long select can be $11 or $12 per hank. Since 1000 feet will cane an area of about four square feet (24 by 24 inches), a hank will serve three to four average chairs, making the cost of the materials $3 to $4 per chair.

Good-quality cane is important and the investment in "select" cane is wise. Poor cane doesn't weave easily because of rough spots. Good cane is smooth and the "eye"—a small lump on the cane from which the leaf grew—is smooth and unbroken.

The materials to cane a chair include the regular weaving cane and binding cane, and the only tools you need are a caning awl and caning pegs. Binding cane is simply strands of cane two sizes wider than the weaving cane, and every hank comes with enough binding cane included in it to take care of several seats.

Caning pegs are tapered hardwood pegs which are pushed into the cane holes in the chair rail to hold the cane in place as you weave. They cost about 60¢ a dozen. However, you don't even have to buy these, since golf tees work quite well in this job. A caning awl can be any long, thin prod, and is used to push the cane through the holes. Cane dealers have these for less than a dollar, but if you already have an ice pick, it will serve just as well.

Cane comes in six widths, and the width you order is determined by the size of the holes in the chair rail. If you have some of the old cane, measure its width. If not, measure the hole size. Consult the chart on this page, and order your cane by both name and hole size.

There is some variation in cane terminology from supplier to supplier. Some call the narrowest strip "carriage" instead of superfine. Your best bet is to send for dealer catalogs before ordering. Most also send along samples.

Some dealers now offer "chair kits." These are complete packages, including pegs, with enough weaving and binding cane for one chair. These kits are fine if you intend to do only one chair, but if you expect to do several, you'll find it cheaper to buy cane by the full hank.

### How to order cane

| Hole size | Distance between holes | Name of cane size |
|---|---|---|
| 1/8-in. | 3/8-in. | Superfine |
| 3/16-in. | 1/2-in. | Fine-fine |
| 3/16-in. | 5/8-in. | Fine |
| 1/4-in. | 3/4-in. | Narrow medium |
| 1/4-in. | 3/4-in. | Medium |
| 5/16-in. | 7/8-in. | Common |

### preparing to weave

The first step is to clean any old cane from the chair. Even if the old cane seat has disintegrated, you may find remnants in the holes, along with a buildup of dirt, wax, and finishing materials. Sometimes the buildup in the holes turns into fairly solid plugs, and if these can't be cleaned with an awl, use your electric drill with a bit of the correct size to do the job.

If the edges of the chair frame which contact the cane are sharp, use a rasp or coarse file to round them off. This will prevent the edge from cutting the cane. Also, see that the edges of the holes are slightly rounded. If they seem sharp, use a small rat-tail file to ease them a little.

If you intend to refinish the chair as well as to recane it, do all the repair and refinishing work before you begin to weave the cane.

Leave the cane in the hank as it was delivered, and pull each strand from the hank as you need it. Select a strand at the looped end of the hank and pull carefully, shaking the hank as you do to prevent tangling.

Make a coil of the strand with about a 6-in. diameter, with the shiny side of the cane on the outside. Use a spring clothespin or other small clamp to hold the coil.

When handling dry cane, either old or new, be careful to avoid a nasty cut. Some strands have sharp edges which can cut like a razor when dry.

Now immerse the coiled cane (2 or 3 coils at a time) in warm water. Some caners add about 1½ ounces of glycerine per pint of water, but this is optional. Use a wide pan or bowl, and allow the cane to soak about 15 minutes, until it becomes soft and pliable.

### beginning the weaving

Put the chair upright in front of you, so that you look down on the seat from the front as you weave. Begin by counting the holes across the back rail. If there is an odd number of holes, put a peg in the center hole. If there is an even number, put a peg in each of the holes at the center. Do the same at the front rail.

continued

**TO REMOVE THE OLD** cane, cut the loops of cane on the bottom with a sharp knife. Be sure you cut every loop completely through.

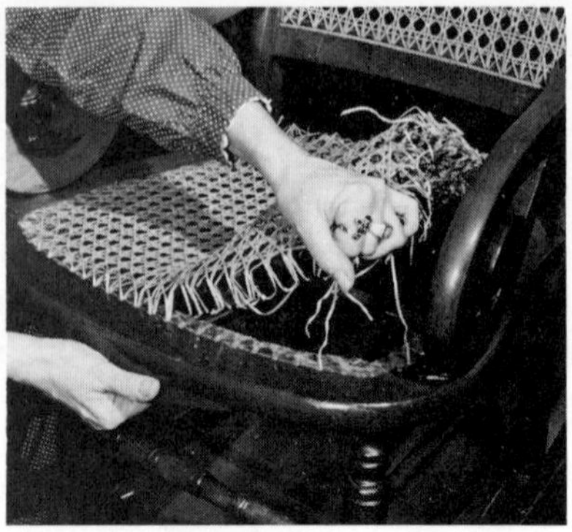

**WHEN ALL LOOPS** are cut, you'll be able to lift off the old cane in one piece. Then clean accumulated dirt from the holes with an awl.

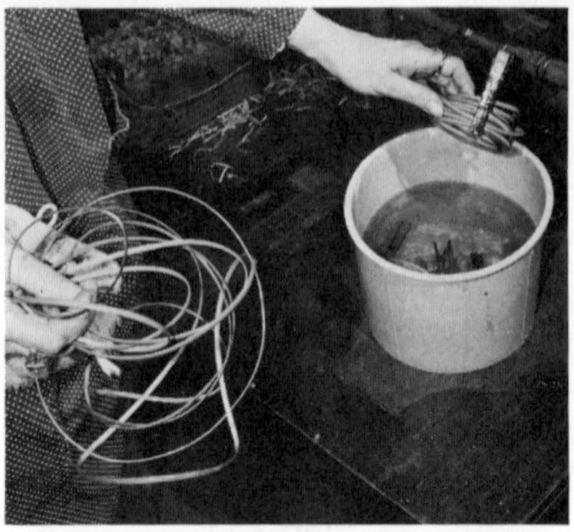

**PREPARE THE CANE** for weaving by coiling it, clamping with a spring clothespin, and soaking in warm water for 15 minutes.

**PUT CANING PEGS** in center holes, then begin at the back and run a strand from back to front. Work from the center to the left.

**WEAVE THE CANE** back and forth, using pegs to hold it when necessary. Start a new strand whenever necessary.

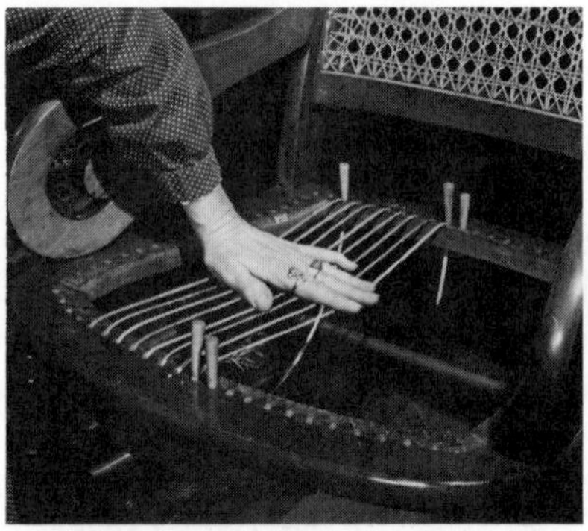

**DON'T WEAVE** the cane too tight, since it will tighten later during the weaving process. Allow a little slack to make subsequent steps easier.

**AT THE SIDES,** cane is not carried all the way to the back holes, but is inserted in side holes. Keep strands parallel.

**THE SECOND COURSE** is woven across the seat, beginning at the back. The strands lie on top of the first course.

**THE THIRD COURSE** is woven front to back in the same holes. Put the new cane to the right of the first strand in each hole.

Now take a coil of cane from the water, remove the clamp, and run the cane between your thumb and forefinger to squeegee off any excess water. Put another coil in the water each time you remove one, so that it will be ready when you need it.

As you put in the first course of cane, you work from the center to the left side, then from the center to the right. Remove the peg in the back rail, insert the cane end about 4 inches, and put the peg back in the hole. Now remove the peg at the front, run the entire strand of cane through it, and reinsert the peg.

Move over one hole at the front and bring the strand of cane up through it. Carry the strand to the back rail and down through the second hole. Continue this procedure until all the holes on the left side have been used, then do the same on the right side.

This is the first course of cane. The second

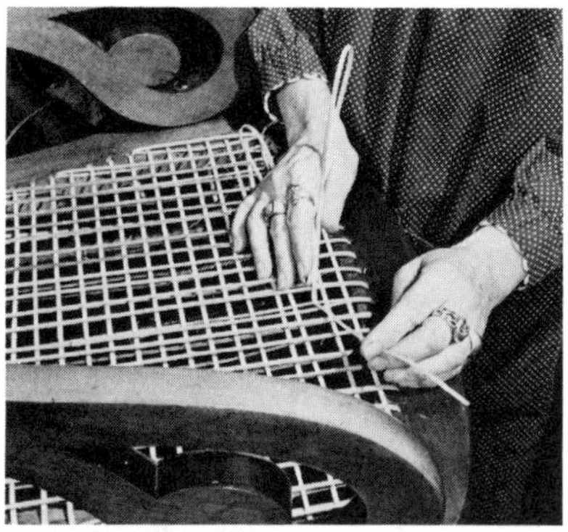

**THE FOURTH COURSE** is woven across the seat, with the new cane going over Course Three and under Course One strands.

**AS YOU COMPLETE** the fourth course, the weaving pattern begins to emerge. Work the fourth course from the front to the back.

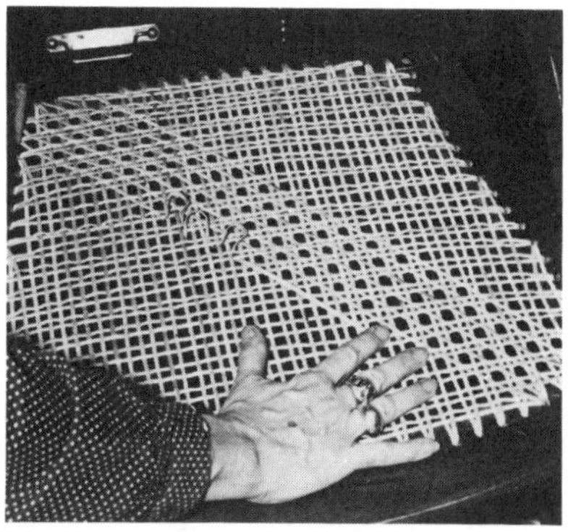

**THE FIRST DIAGONAL** course begins at the left back. Weave it under the side-to-side strands, over the front-to-back strands.

**AS YOU ADD CANE,** holes become crowded. Use your awl or an icepick to help make room in each hole for new cane.

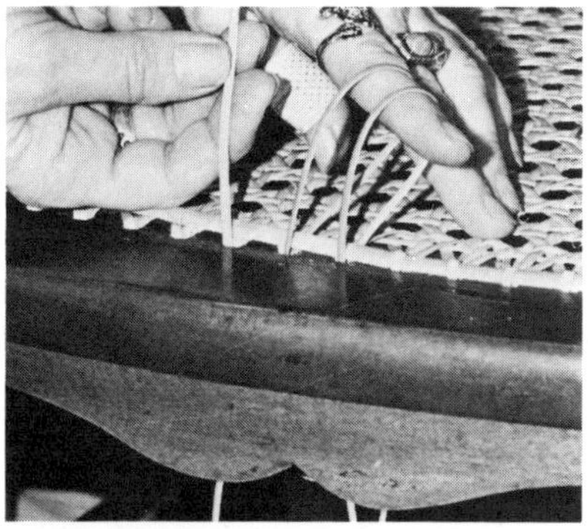

**LAY THE BINDER** cane around the seat edge and secure it by looping regular cane over it at each hole, as shown. This starts the edge trimming.

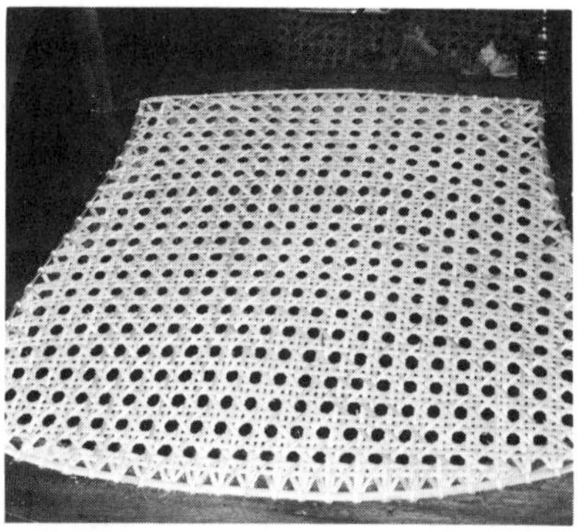

**THE SIZE** of the holes in the completed cane seat is controlled by the number of holes in the chair rail. The more rail holes, the smaller the cane holes.

course runs across the chair. The procedure is the same, except that you begin at the back and work toward the front. These strands lie on top of the first ones. The third course is from front to back again, using the same holes, but laying each new strand to the right of the first strand in the hole.

The fourth course is woven across the seat, with each new strand going *under* Course One and *over* Course Three. The fifth course is diagonal, from the back left to the front right, with the new strands going *over* the front-to-back strands, and *under* the side-to-side strands. The sixth and last course is diagonal in the opposite direction, woven in the same way. Follow the step-by-step process in the photographs, noting how you tie off cane ends and how you install binder cane.

**TO TIE OFF** cane ends under the seat, first run the end under the loop which begins in the hole through which the end comes.

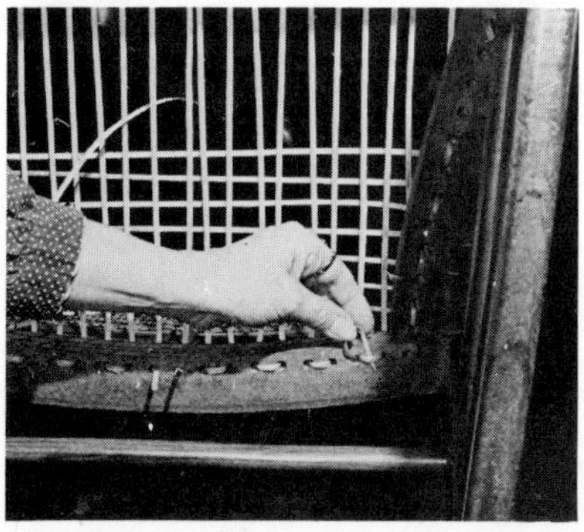

**NOW PULL THE END** back through the loop formed by its own strand. Keep the cane moist and pliable when tying it off to get a neat, small knot.

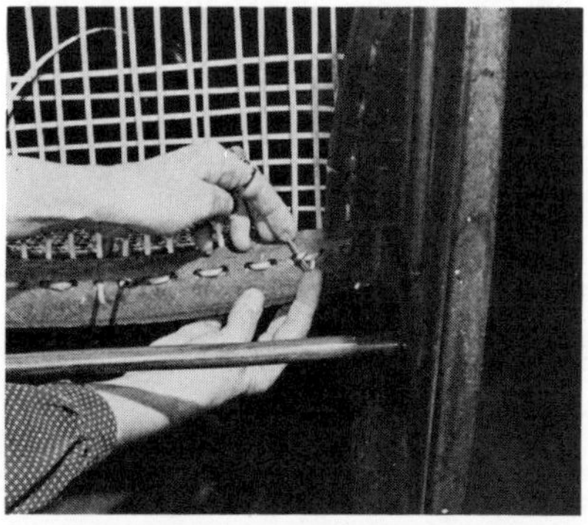

**FINALLY,** pull the tie tight and flat. Making neat tieoffs takes a bit of practice. Trim away excess cane to finish tie.

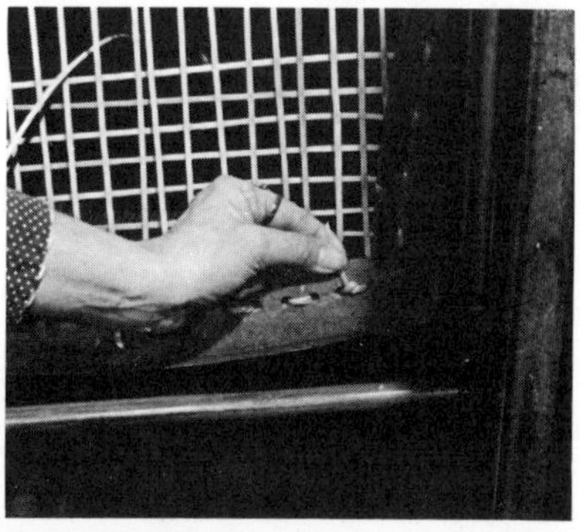

# Prewoven rolls make caning simple

By RALPH S. WILKES

■ TAKE ANOTHER LOOK at that old chair with the hole in its woven seat. See if it has a groove around the edge of the woven portion with a spline in it. If so, machine-woven cane will make the repair easier and less expensive than you might think.

Cane is available in long rolls, 12, 14, 16 and 18 in. wide. A spline is used to wedge it into the seat's groove. Spline comes in diameters of 1/8, 3/16 and 1/4 in., with 3/16 in. the most common.

With household shears cut the material roughly to shape, leaving the recommended allowance on all sides. Soak it in warm water (in sink or bathtub) for a half hour or more so the cane will be pliable. Soak the spline for a few minutes, too. Fit it to the groove and cut it 1/2 in. longer than needed.

Cut about a dozen wedges to hold the cane in the groove. Center the cane on the seat, making sure its glossy side is up and the pattern is not askew. Set the first wedges at front and back; do the same at the sides. Continue this procedure until the caning is taut.

Then use a wedge to force cane into the rest of the groove. Tap lightly with a mallet, being careful not to break the strands.

With a sharp wood chisel, cut off the webbing at the outer edge of the groove and apply glue to the groove. Drive the spline in place using one of the wedges or another small piece of wood.

With a sharp knife, then cut the spline to exact length and drive down the end. The job will look neatest if the ends meet at the center of the seat back.

SEE ALSO

**Adhesives . . . Armchairs . . . Chairs . . . Chests . . . Dining tables . . . Furniture, care of . . . Legs, furniture . . . Sofa-beds . . . Stools . . . Upholstery . . . Wood bending . . . Wood finishes**

**REMOVE** old spline and cane with a wood chisel and clean the groove, being careful not to chip the seat.

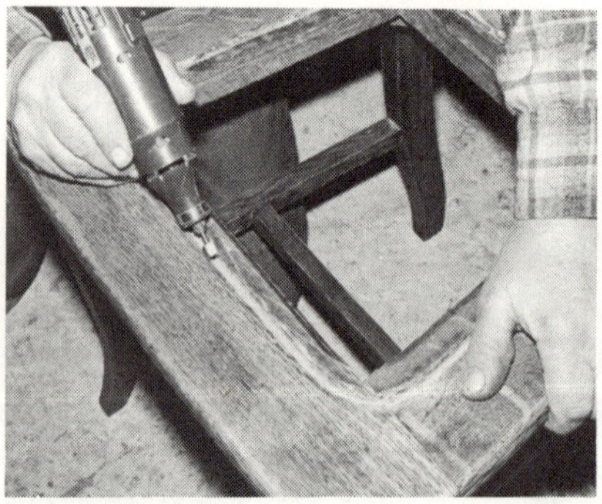

**A HIGH-SPEED** rotary tool is also effective when you're cleaning the old material and glue from groove.

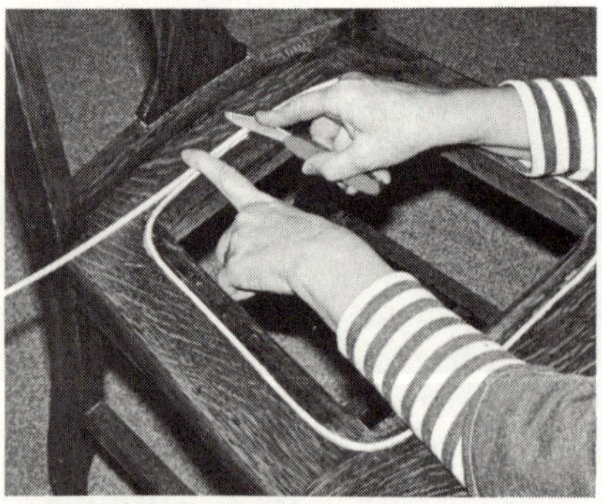

**SOAK SPLINE** for a few minutes in warm water, fit it to the groove and cut slightly longer than you need.

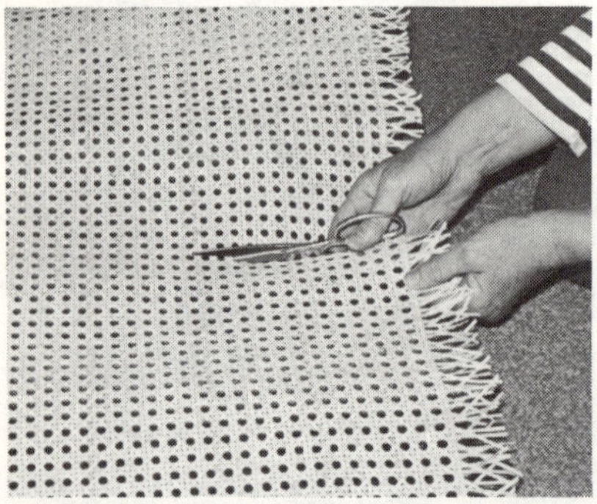

**NEXT,** cut the machine-woven cane about 1 in. larger on all sides than the area enclosed by the groove.

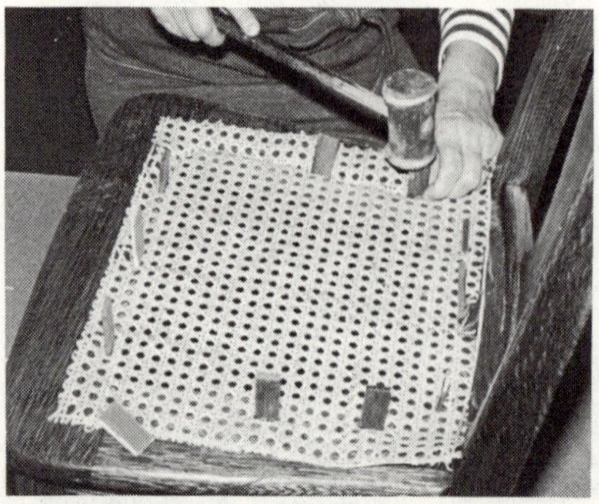

**STRETCH NEW CANE** seat into position using opposing wedges. Use wedge and mallet to tap into groove.

**AFTER USING** a chisel to cut off the surplus cane, run a generous bead of glue in groove, and drive in spline.

### Cleaning gold leaf

*I have a carved American eagle with wingspread of about 20 in. which appears to be covered with gold leaf. It's quite old and I'd like to clean the gold leaf, removing the grime of years. Can I do this without damaging the leaf?—Axel Lundquist, Minneapolis.*

At long range I hesitate to say, but I once cleaned a gilded statuette by applying a mild solution of trisodium phosphate—about a half teaspoonful in a tumblerful of water—with a ball of cotton affixed to a short length of ⅛-in. dowel. I made one applicator for application, another for rinse. With extra care this method proved satisfactory. Press most of the water out of the applicators so that the surfaces won't become overly wetted.

### Growing geranium "tree"

*At a recent flower show I saw what they called a geranium tree. It was 3 ft. tall and blooming profusely. How is this done?—J. Lannon, Dallas*

Geranium trees are quite common in warmer climates. Start with a well-rooted cutting of one of the best flowering varieties. Train it to grow as a single trunk by pinching off all side growth. Stake it early so that the trunk can be supported by plant ties. When it has attained the desired height pinch off the growing tips to induce branching. This procedure takes time and patience, but if you persist your plant will grow into a handsome tree. You can maintain growth with a liquid plant food.

### Bringing out 'tiger-maple' graining

*On many old pieces of "tiger maple" I see the grain is strongly defined, a trick I've never been able to accomplish by staining. How did they do it?—Mrs. John Darlson, Wheeling, W. Va.*

Much of the maple you see in old pieces was cut from mature trees in which the so-called tiger grain was well defined in strong contrasting stripes of more or less uniformity. Material cut from younger trees does not always display this grain strength when finished. Early craftsmen often coated the wood with Van Dyke brown in oil to which a small amount of lampblack was added. This brings out the stripes strongly in contrast when rubbed in with the palm, with any excess immediately wiped off. A later practice is to sear the wood with a torch (a butane torch will do), charring it just enough to accent stripes without blackening lighter portions. This job must be done with care.

### Squeak hunt

*An annoying squeak has developed in my forced-air furnace. I've tried every way I can think of to locate it but I still have the squeak. The noise can be heard upstairs when the blower is operating. It sounds like a metal-to-metal squeak, but where?—D.P., Mich.*

If you're sure it isn't the V-belt (assuming you have a belt-driven blower), the search should center first, on sheetmetal screws which may have loosened and, second, on parts of the fan and motor mounting. Turn the thermostat up until the furnace starts cycling. Then begin the hunt by pressing with a finger at various points over the outside of the sheet-metal jacket. If this doesn't pinpoint the squeak, open the blower compartment and repeat the procedure. Touch all the sheet metal, especially the wall between the blower and exchanger compartments and also the motor mount. Be extremely careful when pressing on parts of the mounting when the unit is running; don't get a shirt sleeve or finger caught in the machinery.

This procedure may take some time, but if you persist, sooner or later you are almost certain to touch the part causing the squeak. The rest is easy; tightening a screw or spraying the troublesome sheetmetal joint lightly with a nonflammable silicone compound usually stops the noise.

### Wobbly casters

*I have a chest (not old) in which the casters are loose and wobbly—two of them drop out when the piece is lifted. Apparently holes for the caster sleeves are too large for any adhesive to hold them in place. Is there any other way of securing the sleeves?—H.R., Md.*

It's hard to suggest what to do without knowing why the sleeves are loose and by how much. Small wedges pared to a thickness that would give the sleeves a tight fit when driven home in glue would be effective, but the problem is to determine just the right thickness to give a tight fit without splitting the post or leg into which they are driven.

Perhaps it would be better to bore out the existing holes to take a length of hardwood dowel having a diameter greater than the hole required to take the sleeve. Tap the dowel into place with glue in the hole and, when the glue is dry, bore a hole to take the sleeve in a moderately tight fit. Before driving each dowel, flatten one side slightly to permit escape of entrapped air and also allow the glue to work up alongside the dowel and anchor it securely.

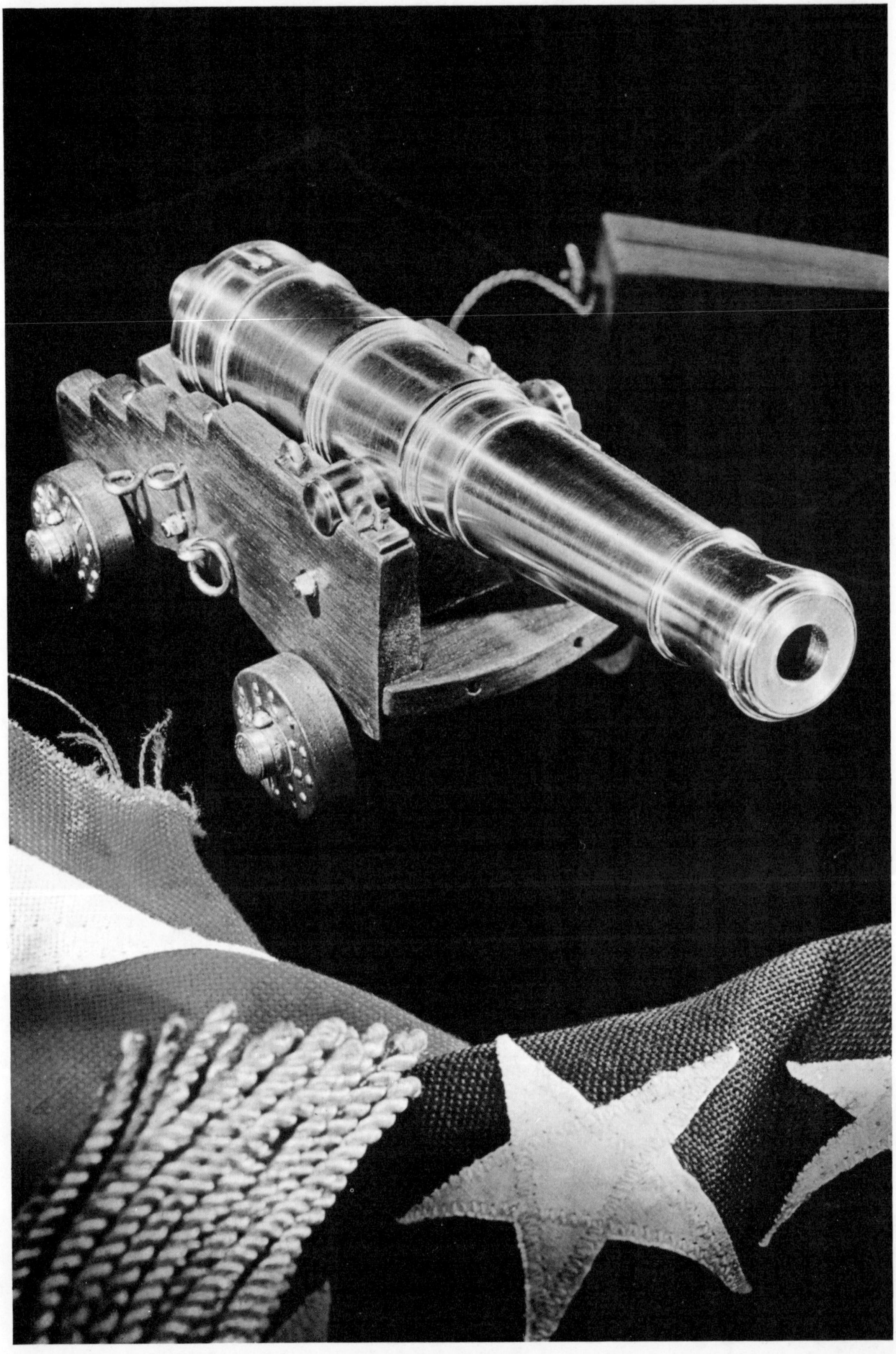

# Model this 24-pounder

By WALTER E. BURTON

**1/20th the actual size of a cannon from the main battery of the U.S. frigate *Constitution*, "Old Ironsides," this gleaming replica makes an attractive ornament for desk or bookshelf. Here are detailed plans and directions so that you can make it yourself**

■ IF YOU ARE a dyed-in-the-wool modelmaker, you know that plastic models can't begin to give you the satisfaction and pride of accomplishment you get from building an authentic replica from scratch. You'll get such a charge from modeling and displaying this 24-pounder from the main battery of the U.S. frigate *Constitution*.

While both English and American types of 24-pounders were in *Old Ironside's* complement of guns, this model, in gleaming brass and contrasting walnut, is patterned after the 12 American-type guns which were deployed at the forward and after ports. It's based on original shipyard plans which were followed in the 1927-1931 restoration of the *Constitution*.

The original guns measured 9 ft., 5¾ in. in length and were of gray cast iron. The wooden carriages were of white oak, and most of the fittings were steel. The barrel of our model is brass and is approximately 5½ in. long, which makes it about 1/20th actual size.

Carriages for the guns in the main battery varied in size according to the gun deck, and the heights of the carriage bed blocks, which were used for elevating the barrel, were determined by trial during firing missions.

SEE ALSO

**Airplane models . . . Jeweler's saws . . . Metal casting . . . Metal-lathe techniques . . . Modelmaking . . . Ship models . . . Tools, hand**

**BARREL BLANK**, mounted between centers, is driven by faceplate dog made from a ball-bearing ring.

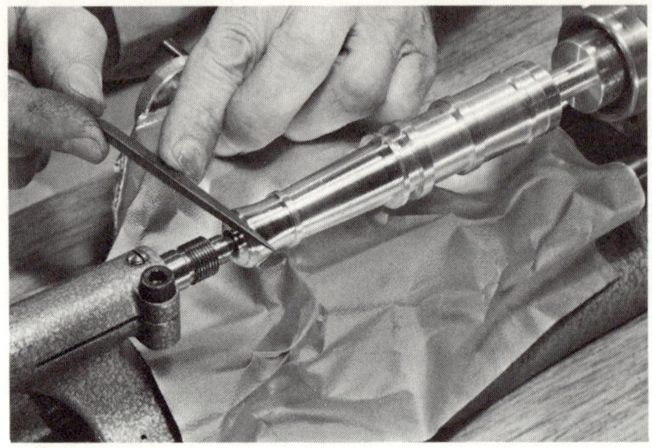

**AFTER RIDGES** are machined square, they're rounded with Swiss-type file as done here at end of muzzle.

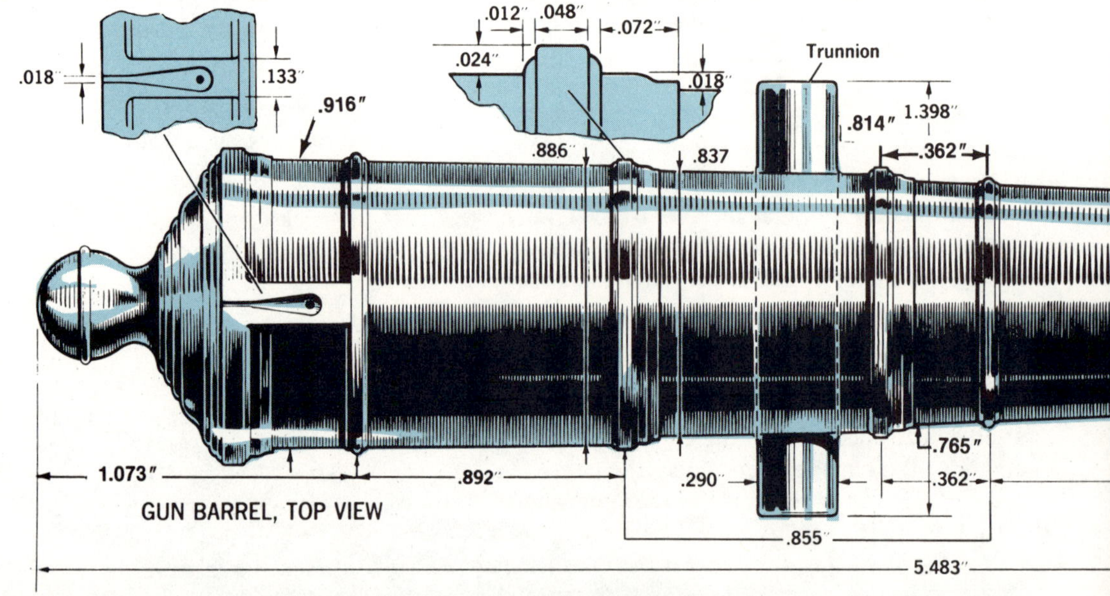


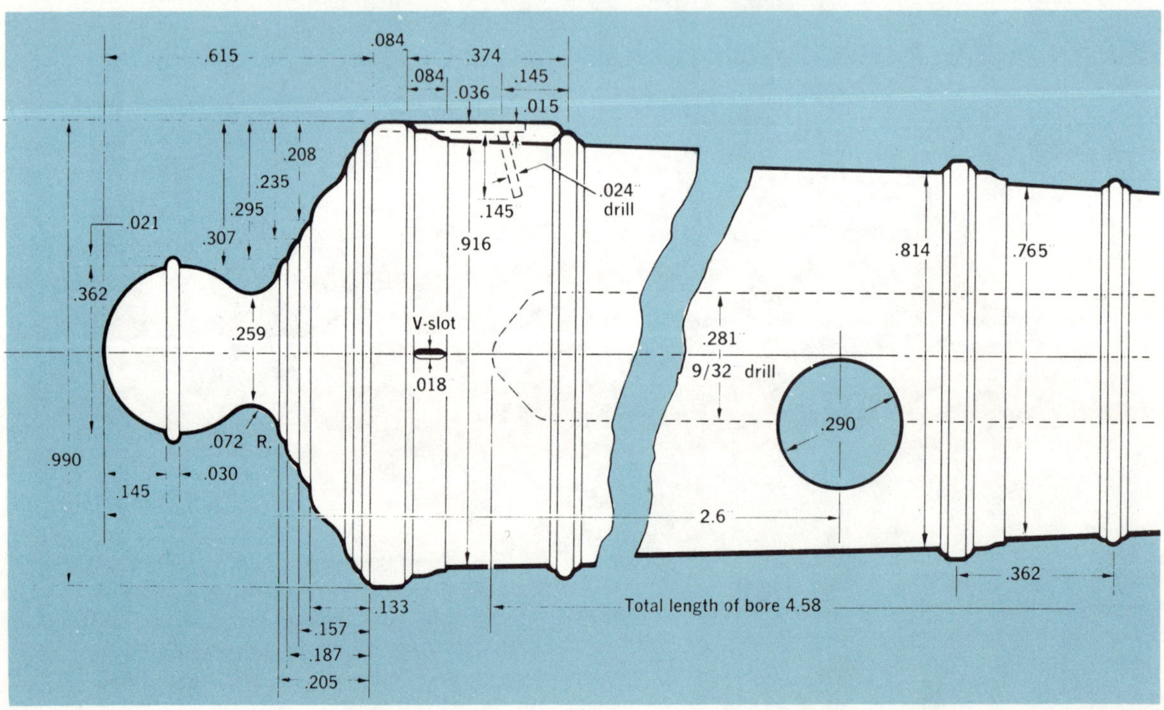

**HAND-HELD CHISEL** resting on blank bit clamped in toolholder is used to shape final breech contours.

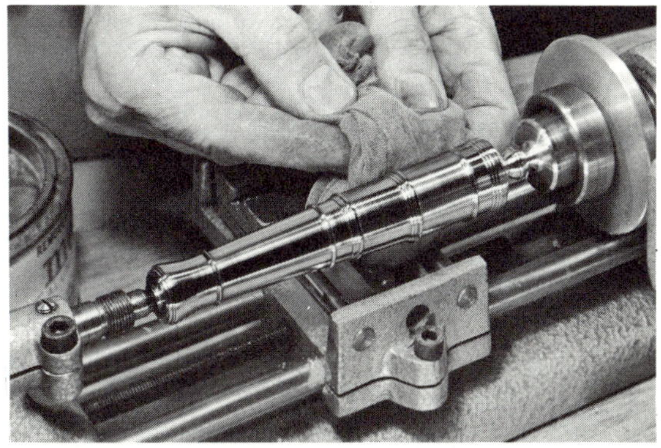

**COMPLETED BARREL** is polished in the lathe with a brass polish; ball end is hand-finished later.

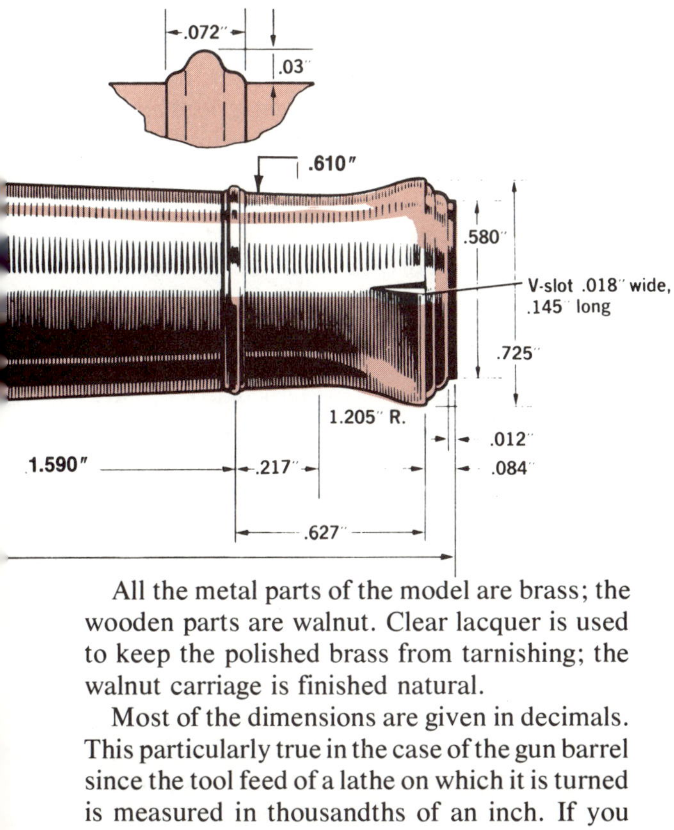


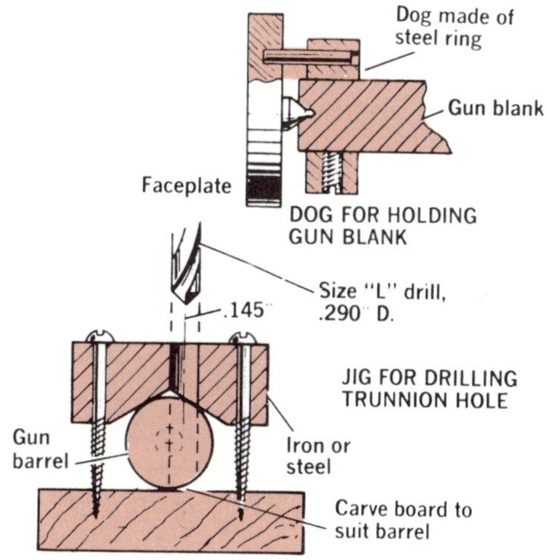


**JIG IS USED** to drill hole through barrel for trunnion rod. Masking tape protects the barrel.

All the metal parts of the model are brass; the wooden parts are walnut. Clear lacquer is used to keep the polished brass from tarnishing; the walnut carriage is finished natural.

Most of the dimensions are given in decimals. This particularly true in the case of the gun barrel since the tool feed of a lathe on which it is turned is measured in thousandths of an inch. If you prefer to work in fractions, you can convert the measurements with the aid of a decimal-equivalent table.

The gun-barrel blank is a 1-in. brass rod about 7 in. long. This length was selected as a convenient size for the Unimat lathe which I used in making the complete pilot model. Because a 3-jaw chuck would cut down on the usable length, the rod was center-drilled and mounted between centers, and driven by a dog made from the outer ring of a ball bearing having a 1-in. bore. However, any similar ring will do. If your lathe lacks bed length to drill the center holes with a bit in a tailstock chuck, lay out centers carefully and use a center drill in a drill press.

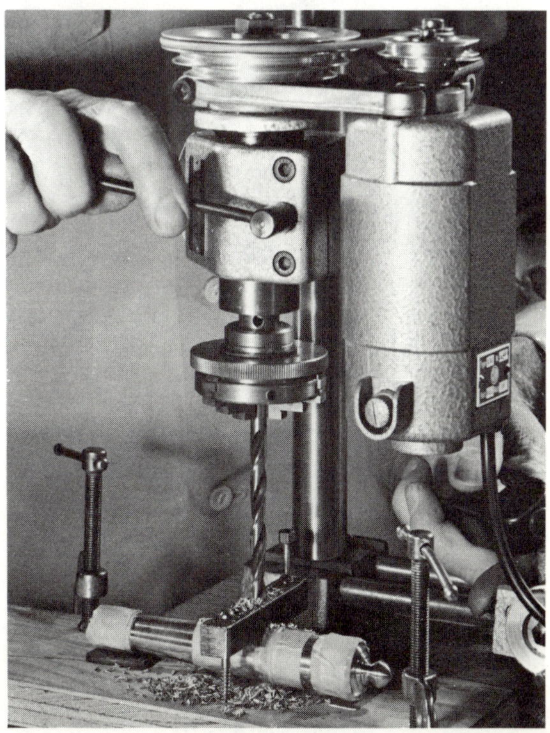

**THIN MILLING CUTTER**, on head spindle arbor, cuts the walnut side members of the gun carriage.

**CUTS AT RIGHT ANGLE** to the first ones are made in both side pieces at the same time; finish by hand.

**CARRIAGE TRUCKS** are turned to their final diameter by mounting discs on arbor of 5/16-in. bolt and two nuts.

**HOMEMADE "PLUG CUTTER"** from 1/8-in. pipe is used in drill press to round ends of wooden axle-tree parts.

When machining the barrel, follow the drawings on these pages. To form the rounded bands and contours, machine them with square edges, then round them with a fine file or a hand-held chisel.

Trunnions are usually an integral part of a real cannon. In the model barrel, however, they are cut from brass rod. To drill the hole in the barrel for this rod, use the jig detailed on page 619. If you lack an L-size drill (.290-in. dia.) you can make a 9/32 or 19/64-in. hole and match the trunnion notches in the carriage. The rod forming the trunnions can be a drive fit or held with a setscrew on the underside of the barrel.

To bore out the barrel, use a 9/32-in. twist drill, starting at the center hole in the muzzle end. The drawing shows the bore full length, as in the original cannon, but in the model it need extend only to the trunnion hole, enough to keep the barrel from being muzzle-heavy.

The powder-igniting arrangement on top of the breech is formed as shown. The teardrop recess is a separate piece of brass soldered or cemented with epoxy to the barrel. V-shaped sighting notches on top and each side of the barrel can be hand-filed or scribed while the barrel is in the lathe. To do this, move the lathe carriage back and forth and feed a pointed tool so it shaves a notch .018-in. wide.

Before removing the barrel, polish it with very fine abrasive cloth. Then use a brass polish for a final luster. After cutting scrap from the breech carefully smooth and polish the rounded end. If you solder the powder "pan" to the barrel, repolish the metal to remove heat discoloration.

Full-size patterns for wooden parts on the gun carriage are shown at the right. As for the sides, made in pairs, you'll find it helpful to make a master pattern from aluminum or cardboard, then drill the ring and tie-bolt holes in it. Saw

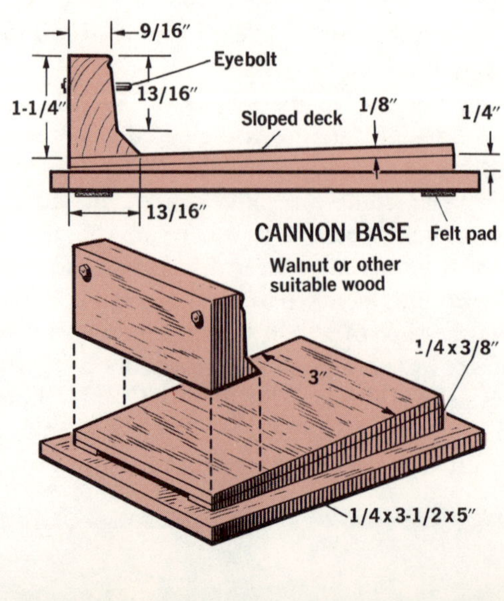

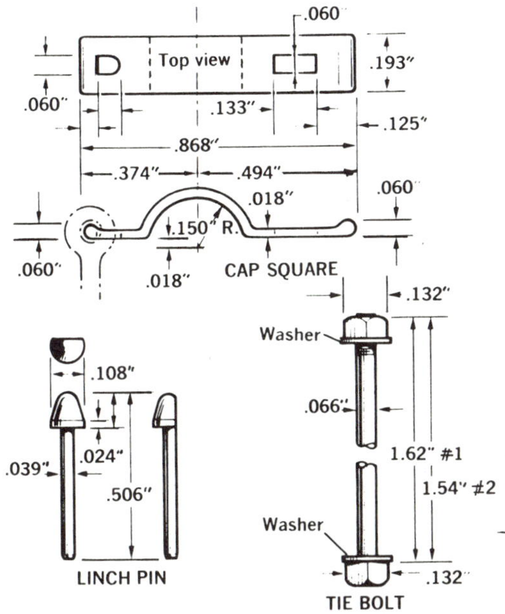


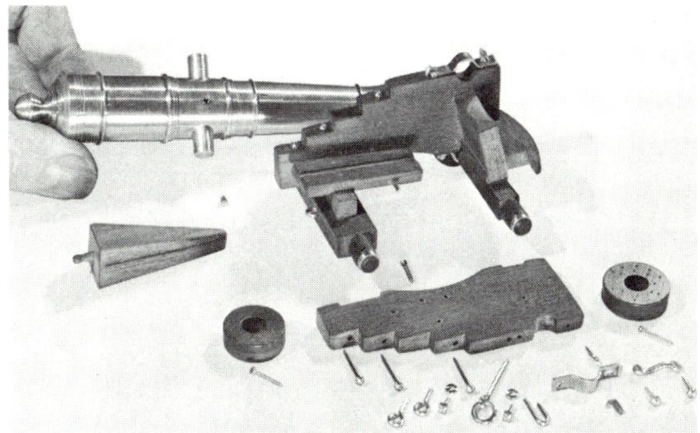

**PARTLY ASSEMBLED CARRIAGE**: Eyebolts and other parts are half-length but seem to pass clear through.

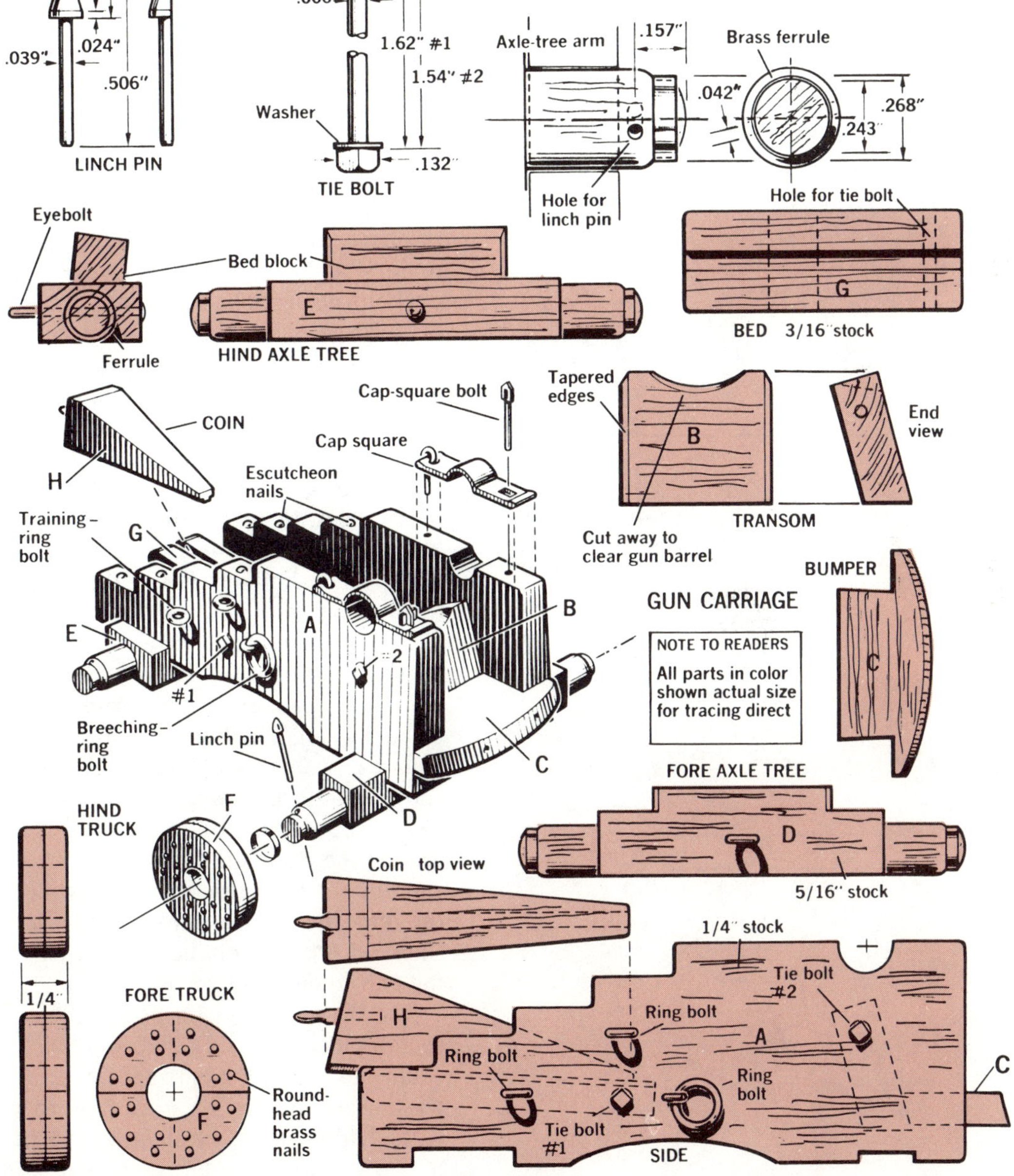


NOTE TO READERS

All parts in color shown actual size for tracing direct

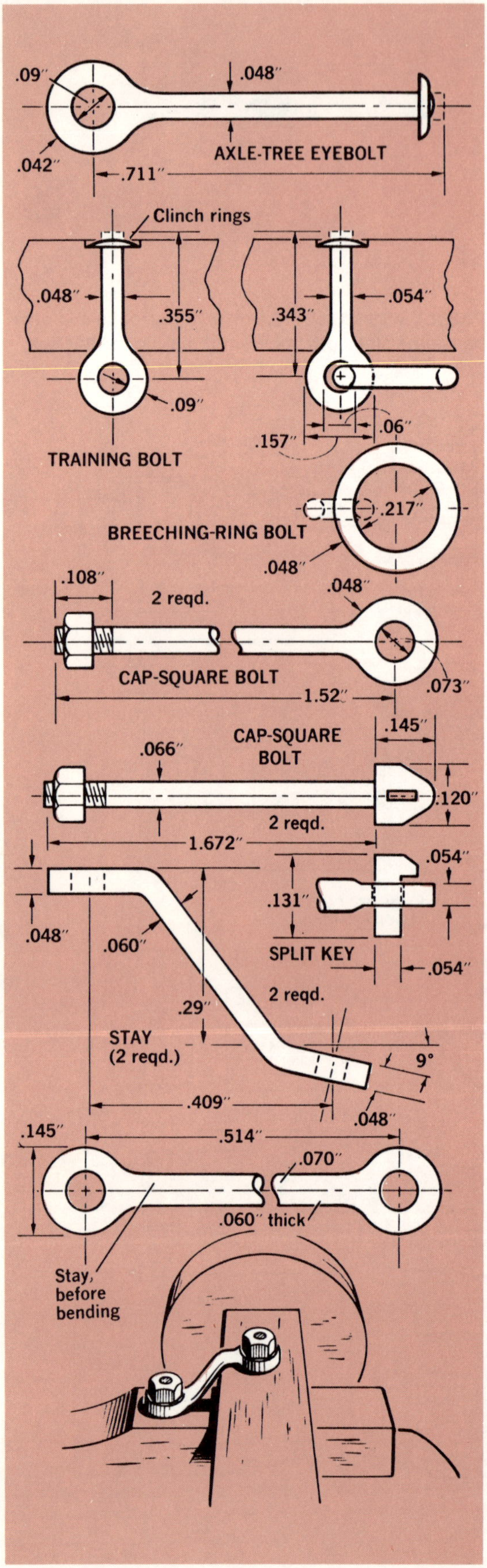


both sides at one time, tape the template to them, then drill the holes slightly smaller than the eyebolts. Note that the sides taper upward slightly from the base to top.

Carriage parts are shaped from black walnut and glued in the order shown. The trucks of the original cannon were built up in layers, as detailed for the model, but you can cut them from ¼-in. stock and represent the joints with knife cuts. Note that hind trucks are smaller than the fore trucks. Use tiny brass escutcheon pins for rivets.

Round ends of the fore and hind axle trees can be "turned" to fit the trucks with a homemade plug cutter chucked in the drill press. It's made from ⅛-in. pipe reamed with a 5/16-in. twist drill. Form teeth around one end with a file. Axletree ferrules are turned from brass rod and bored to form a press-fit ring.

Tie bolts and cap-square bolts are shown full length, but you can simulate them merely by inserting the "heads" of the bolts in shallow undersize holes. Likewise, the eyes of the eyebolts need not be solid rings, as shown. Simply form an eye with needle-nose pliers and force it into the wood so you can't see the open end. If you can find actual bolts and nuts from junked timepieces, your model will look more authentic.

Cut stays for the fore axle tree from .060-in. brass sheet, then file, drill and bend to shape. Shape the linch pins from fair-size brass escutcheon pins, filing a flat on their heads to clear the axle trucks.

Cap squares that clamp over the trunnions can be shaped from .060-in. brass. File away the middle portion to .024-in. thick, then cut the piece in half lengthwise to form the two cap squares. Round the thick ends remaining as shown and form the slots by drilling and filing. Brass should be soft enough to bend easily around the trunnions. Note that one end of the cap squares pivots on an eyebolt while the slot in the opposite end engages a bolt with a slotted "head." A brass split key fits the slot and locks the cap square around the trunnion.

The base detailed on page 620 represents a section of sloping deck and ship rail. Make it of walnut to match the carriage or of lighter, contrasting wood. For a darker tone than you get with clear lacquer, rub in boiled linseed oil and let dry a day or two. Or, seal the first or second application with thinned shellac (after oil has dried), sand lightly and apply a final lacquer coat.

# How to outfit for a wilderness canoe trip

By NICK KARAS

**The way you outfit for a canoe trip into the wilderness will depend on the purpose and length of your trip, the time of year and your individual needs for comfort. Here are some of the basics for any trip**

ONE MAN'S WILDERNESS is another man's back yard, and how to outfit or prepare for an excursion into either does and can differ considerably. There's no such thing as a standard outfitting list or technique because each trip will vary, based on the number in the party, the purpose of the trip, the kind of water to be traveled upon, the duration of the rides and the time of year the canoes go in the water.

The best approach might then be to set up a theoretical trip and let you modify it according to your needs. The following list of materials, then, is for two men canoeing in semi-wilderness country, that is, they can drive to a putting-in place and take-out location. The river they run will have a few portages, maybe a lake or two, some fast, but little white water.

I like the early fall for trips into the bush because black flies and mosquitos are usually at a low ebb, the weather is cooler and more pleasant than mid-summer and the trees are likely to be in better color.

How to canoe is another story; this one is concerned with what you should bring along to make it successful. "What" will depend on the things you insist on having to be comfortable away from home. Tolerances differ with the individual.

Before you start a trip into the bush, you should make a checklist of what you think you'll need. If you've done it before, you'll realize that you have several checklists, probably these six: a basic outfitting list, a basic food list, a clothing list, a ditty-bag or personal list, a list of options concerned with the primary purpose of the trip if it is more than just canoeing, and a medical list that can be cut to one if you pick up a complete first-aid kit.

Believe it or not, all this equipment will fit nicely into a 16-foot canoe with enough room at the gunwales for ample safety. The small outboard motor is on the basic outfitting list for several reasons. In case of an emergency, you need it to get out in a hurry. If your trip doesn't end at a take-out place, it can get you upstream. It comes in handy for side trips up feeder streams and saves your time crossing lakes and on the open water.

Most equipment listed is self-explanatory. Unless you're taking along an extra large canoe, you can cut out some of the items. Remember, it all goes on your back each time there's a portage. The pop-type tent means you won't have to bring along poles or cut supports. An Adirondack packbasket is great because it's self-supporting and has canvas covers for use when it rains. Seat cushions are important to soften the load during the trip, serve as pillows and take the bumps out

SEE ALSO

**Backpacking . . . Boat camping . . .
Boating, safety . . . Boating, white-water . . .
Camping . . . Hunting . . . Life preservers . . .
Map reading . . . Outboard motors**

## CANOE-CAMPING EQUIPMENT LIST

1-½ hp outboard motor
Seat and boat cushion
Outboard gasoline, oil mixed
Catalytic heater
Three paddles
Collapsible anchor
Cooking utensils
Cooking stove
Folding saw
Frying pan
Insect repellent
Rucksack
Fishing and toolbox
Grate
Pack-in fishing rods
Anchor and bow lines
Poncho
Adirondack packbasket
Sleeping bags
Outboard bracket
Lantern
Ax
Food box, heavy plastic with snapping lid
Canoe carrying yoke
Pop Tent
Folding shovel

## BASIC OUTFITTING LIST

| | |
|---|---|
| Tent, ground cloth | Scouring pads |
| Sleeping bag(s) | Dish towel |
| Air mattress | Poncho, rain suit |
| Catalytic stove | Life jacket(s) |
| Fuel: stove, lantern | Seat cushion(s) |
| Lantern mantles | Paddles (3) |
| Flashlight | Carrying yoke |
| Spare batteries, bulbs | Anchor, anchor line |
| Cooking stove | Bow lines |
| Cooking utensils | Tie-down lines |
| Eating utensils | Canoe patch kit |
| Cooking grate | Plastic tape |
| Packsack | Stovepipe wire |
| Ax, folding saw | Outboard motor, 1½-hp |
| Folding shovel | Motor bracket |
| Insect bomb | Motor fuel |
| Stick matches | Topographical maps |
| Liquid soap | Compass |

## BASIC FOOD LIST

Food lists depend on your personal taste, no pun intended. The following one covers only the basic staples, to which you can add the real bulk of the list—after planning meals, three a day, for every day you'll be away.

| | |
|---|---|
| Shortening, flour | Evaporated milk |
| Margarine | Coffee, tea |
| Salt, pepper | Powdered chocolate |
| Sugar | Catsup |
| Dried milk | Mustard |

## CLOTHING LIST

This list will depend greatly on the time of year you make the trip and how often you like to change clothing.

| | |
|---|---|
| Windbreaker | Handkerchiefs |
| Light ski jacket | Hat, with visor |
| Long-sleeved shirts | Gloves: cotton |
| Undershirts | Socks: wool, cotton |
| Underpants | Shoes: field, tennis |
| Trousers, shorts | Heavy belt |

## DITTY-BAG LIST

| | |
|---|---|
| Toothpaste | Bath towel |
| Toothbrush | Sewing kit |
| Chapstick | Sunglasses |
| Pocket knife | Bar soap |
| Sheath knife (filet) | Paper towels |
| Insect repellent | Toilet tissue |

## FIRST-AID KIT

| | |
|---|---|
| Adhesive tape | Merthiolate |
| Antiseptic ointment | Needle, tweezers |
| Adhesive bandages | Gauze pads, strips |
| Burn ointment | Salt tablets |
| Aspirin | Halazone tablets |

of rocks and logs around the campfire at night, as well as being lifesaving devices. Even so, life jackets should also be in the canoe and worn when the water turns white or a long crossing of open water is at hand.

When canoeing in the rain, what to wear is a toss-up. Many older canoeists prefer a poncho; it gives them arm freedom and can be used to cover other gear when you're not under it. However, some of the new rainsuits are equally comfortable and can be stored in minimum space. The top also acts as a windbreaker on cool days.

If you don't take a fuel stove, then a fire grate is indispensable. Even if you have a stove, it's handy if the fuel should run out or you need an extra burner or two. It can be quickly set up between stones or logs.

If two paddles are good, then three are better to take on a trip. They have a way of breaking; the spare should always stay lashed inside the canoe in case you lose a paddle in a spill. It's too difficult to cut a new one in the bush that is as light or efficient as the extra one you bring along.

Spare batteries for a flashlight are fairly standard in most outfitting kits, but be sure to add a few bulbs. They do burn out and take such little space. Canoe repair kits—aluminum especially—can consist of heavy, reinforced tapes or tubes of liquid metal. A rubber patch kit for your sleeping bag or waders is also a good item to throw into the bag.

**use waterproof sleeping bags**

Sleeping and clothing bags should be of waterproof plastic or rubberized cloth. There's nothing worse after a hard day on the paddle than a wet sleeping bag when it didn't even rain.

Electrician's plastic tape and stovepipe wire will have endless uses. A pair of pliers and a screwdriver should be in your tackle box. They're handy in so many unpredictable ways.

A Citizens Band radio is rather limited in range, with a maximum up to 20 miles. But sometimes that might be all the distance you need for an answer to your call for help. At night, the bounce can carry your plea halfway across the country. A portable radio is also a good idea. In some heavily canoed areas, radio stations have special programs for canoeists with messages from the outside world.

You might not find all the items on the lists helpful, or you might want to make additions to it to suit your own needs. The list may have only given you a few ideas, but it's a place to start to build your own outfitting guide.

**YOU CAN WEIGH** your own letters by using a round pencil and 12-inch ruler, plus coins that equal one ounce (five nickels and a penny are within a few grains). Center the ruler on the pencil and place coins over the 10-in. mark. Then center your letter over the 2-in. mark. If it tips the "scale," it weighs over an ounce.–*William Slamer, Menomonee Falls, Wi.*

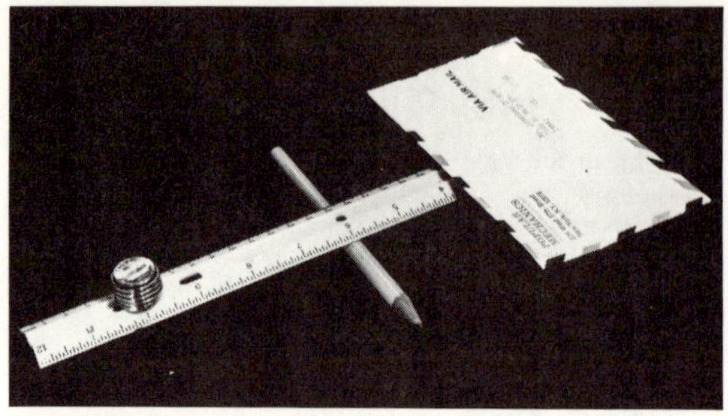

**IF YOU INSTALL** a sturdy hose clamp on a drillpress column below the table, it will limit table movement downward. It also serves as a stop in case the table is accidentally loosened enough so it can drop. The clamp shown is a worm-gear type and is easily loosened and tightened whenever you want to adjust the clamp's position.–*Walter E. Burton, Akron, Oh.*

**ANY TIME** you have to park your car in direct sunlight for an extended time, you're sure to return to a "hot seat." But if you take time to cover the car seats with aluminum foil, much of the sun's heat is reflected away and the surface below stays cooler. Foil-surfaced paper shown came from packages of photographic paper.–*W. B. May, Oak Park, Il.*

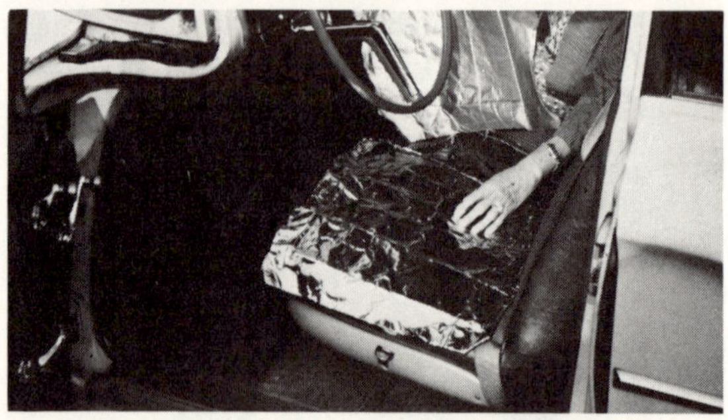

**SOMETIMES A RUBBER STAMP** will make a better impression if the material being stamped is placed on a resilient support such as several thicknesses of a handkerchief. For a more permanent cushion, apply one or more layers of self-adhering felt to the inkpad cover as shown. Work to be stamped is then laid on the closed box.–*B. W. Ervin, Kent, Oh.*

**HERE'S A FAST,** simple way to be sure your drillpress table is at a right angle to the spindle. Replace the tool bit in an adjustable circle cutter with a sharp pencil. Then, bring the spindle down until the point barely touches the table. Revolve spindle by hand until pencil grazes table evenly through a full circle.–*Bert Slanhoff, Scarsdale, N.Y.*

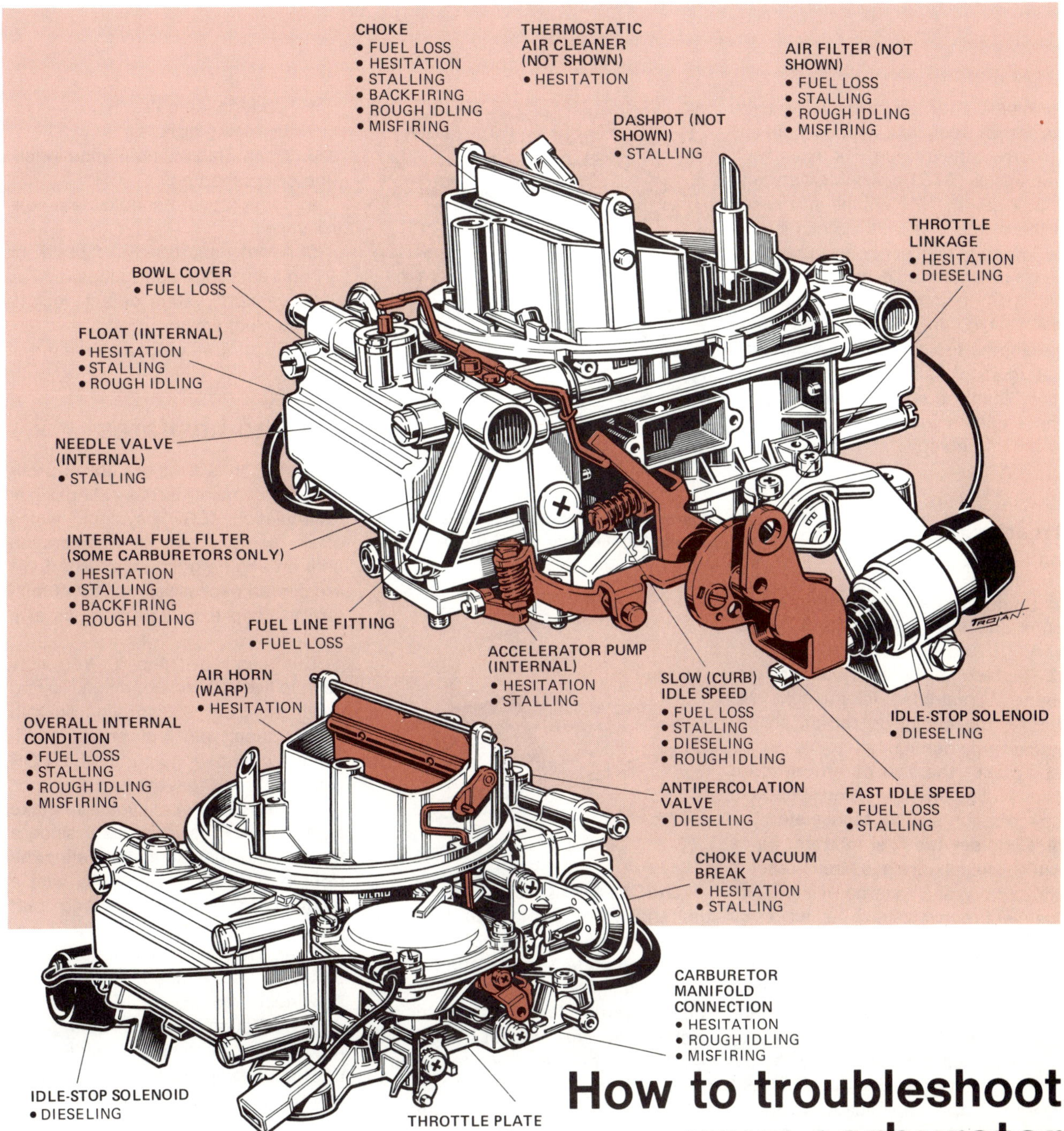


# How to troubleshoot your carburetor

■ MORE CAR OWNERS are concerned about getting better gas economy than any other carburetor-related problem. Check the following:

**1.** In the absence of a mechanical problem, your engine will consume a certain amount of gasoline. As long as you operate the car sensibly, you are getting the best mileage.

**2.** More than any other factor, thoughtless operation causes poor fuel economy. Excessive speed, prolonged warm-up, and jackrabbit stops and starts waste tons of gas in this country every day.

**3.** Gas mileage performance figures posted by new-car manufacturers are estimates attained in the laboratory under fixed conditions. No allowance can be made for variations in driving technique, weather, road or mechanical conditions.

**4.** If a manufacturer finds a condition in a new car affecting fuel per-

SEE ALSO

**Chokes, auto . . . Dieseling, auto . . . Differentials, auto . . . Distributors, auto . . . Engines, auto . . . Fuel systems, auto . . . Misfiring, auto . . . Stalling, auto . . . Tune-up, auto**

formance after the car is in the field, he issues technical service bulletins to correct the trouble. For example, if you own a 1975 Ford or Mercury with a 351-cu.-in. "M" engine and asked if there was a service bulletin issued to improve fuel economy, your service manager would uncover bulletin No. 107. Instructions state that certain 351-cu.-in. "M" engines would give better fuel performance if the initial ignition timing is advanced 4° more than the specification given on the engine's emission-control decal in the engine compartment.

### Troubleshooting bad mileage

■ **Look for fuel leaks.** Tighten fittings and replace gaskets if necessary.

■ **Replace dirty air cleaner element.** A restricted filter blocks air, forcing the engine to run on an unnecessarily rich mixture.

■ **Adjust the carburetor's** curb (slow) idling and fast idling speeds to specification using a tachometer.

■ **Consider the fuel mixture.** Since 1968, idle mixture has been preset at the factory and external mixture adjustment screws have either been capped to limit travel or eliminated altogether. Mixture will normally stay constant until the carburetor gets dirty enough to need an overhaul. This condition can be determined by checking carbon monoxide output with a CO tester. If output exceeds manufacturer specifications and a regular engine tune-up fails to bring CO into line, the carburetor should be overhauled. This will improve fuel economy.

■ **A misadjusted or malfunctioning automatic choke** is another reason for excessive fuel consumption. Make sure the choke control is adjusted to manufacturer specification. Warm up the engine and remove the carburetor air cleaner. Is the choke plate wide open?

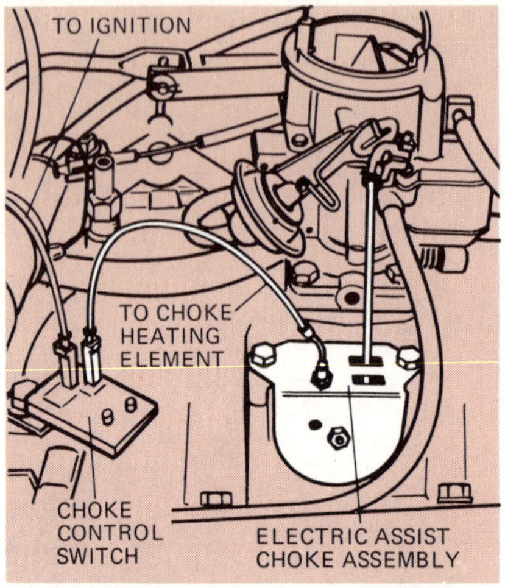


**Installation of typical electric-assisted choke (above) and detail (below).**

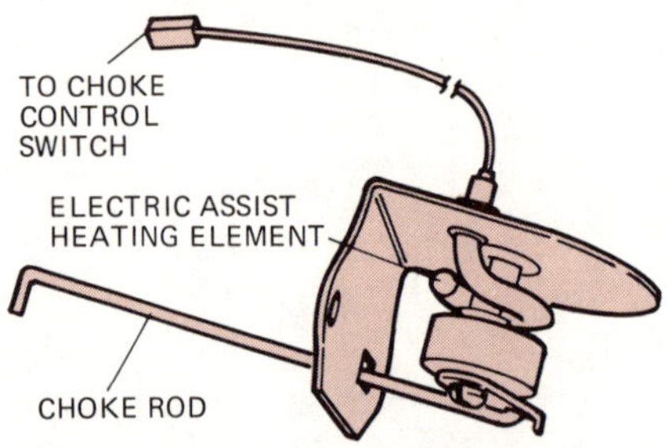


***QUICK FIX:*** Clean linkage and butterfly valve pivots with choke or carburetor cleaner. Do not oil.

If cleaning fails to overcome choke plate sticking, the choke system needs work. With nonelectric-assisted stovetype chokes (the choke control mechanism is on the carburetor), replace the thermostatic spring, check the heat tube for damage and free up the choke piston.

With nonelectric well-type chokes (the choke control is in a well in the exhaust manifold), replace the entire control.

With electric-assist chokes, check wire connections. Then test the control switch and electric heating element.

The non-carburetor-related mechanical problems that produce low gas mileage are the following:

**1.** Underinflated tire.

**2.** Dragging brakes.

**3.** Restricted exhaust (mud in tailpipe or crushed pipe).

**4.** Inoperative manifold heat-control valve.

**5.** Worn, improperly gapped or wrong heat range sparkplugs.

**6.** Ignition system defect, including incorrect timing.

**7.** Sticking valves, camshaft out of time.

## He who hesitates

Hesitation is a major current problem. One reason is the exhaust gas recirculation (EGR) system, which often disrupts smooth acceleration until an engine gets warm. The EGR creates an excessively lean fuel mixture to keep exhaust emissions at a low level.

This can't be helped. You may have to live with the condition. To find out if the EGR is disrupting smooth acceleration with the engine cold, clamp the hose going to the EGR valve and road-test the car.

If hesitation is eliminated, make sure the EGR valve is clean since a fouled valve may be accentuating the trouble. Tap the valve lightly with a hammer to free carbon and buff mounting surfaces with a wire brush. The valve has to be dismounted if this servicing is required. Inspect the EGR hose, too.

If hesitation prevails during the road test, something besides the EGR is causing the trouble. The reason for hesitation is usually carburetor-related.

### Troubleshooting hesitation

■ **Clean throttle linkage** and check it for binding.

■ **Perhaps throttle plates are bent.** This is unlikely unless the carburetor has been recently overhauled and plates were accidentally damaged.

■ **With the choke plate open,** look into the carburetor throat as the accelerator pedal is rapidly depressed. A steady stream of gas should discharge into the venturi. If not, disassemble the carburetor and inspect the accelerator pump plunger (diaphragm) for deterioration. A faulty accelerator pump plunger is the main cause of engine hesitation on acceleration. Replace.

■ **The carburetor's high-speed** (main metering) circuit may be clogged. You can get a good idea by idling the engine at about 1000 rpm and partially covering the air horn with your hand. This should cause more fuel to be discharged from the high-speed circuit, causing an increase in engine speed. No increase? Carburetor cleaning is called for.

■ **Proper carburetor float level** position is important. If it's too low, the metering circuits won't be able to deliver sufficient fuel; if too high, flooding will occur.

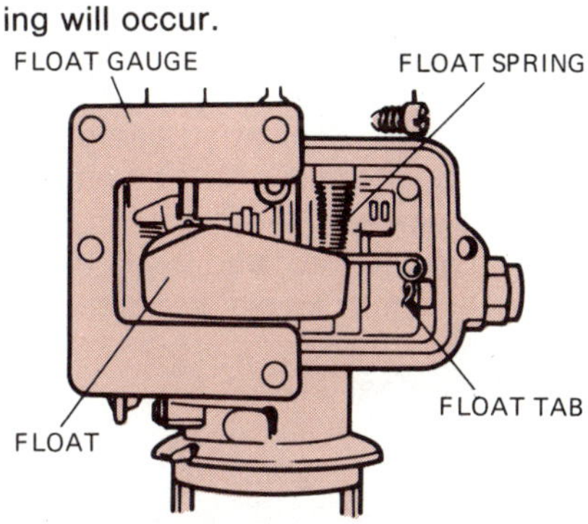


**Float level is adjusted by bending tab. Carb is inverted to show parts.**

***QUICK FIX:*** If a float springs a leak and takes in gas, the float level will be consistently low. Hold float to your ear and shake. Sloshing? Discard the float.

■ **A fuel filter** that's partially clogged causes hesitation. Remove an in-line filter from the fuel line. Remove an internal filter from the carburetor inlet.

■ **A malfunctioning choke** could cause hesitation as well as contribute to poor fuel economy (see earlier text).

■ **Chokes in some cars** are controlled by a vacuum break that opens the choke plate partially when a cold engine is started. This keeps the engine from loading up with gas. If the vacuum break is not properly adjusted, the choke plate may not open enough or may open too much. Hesitation will result. The vacuum break is adjusted by bending the vacuum break rod until a specific size gauge (see manufacturer specs) fits snugly between the choke plate and carburetor throat.

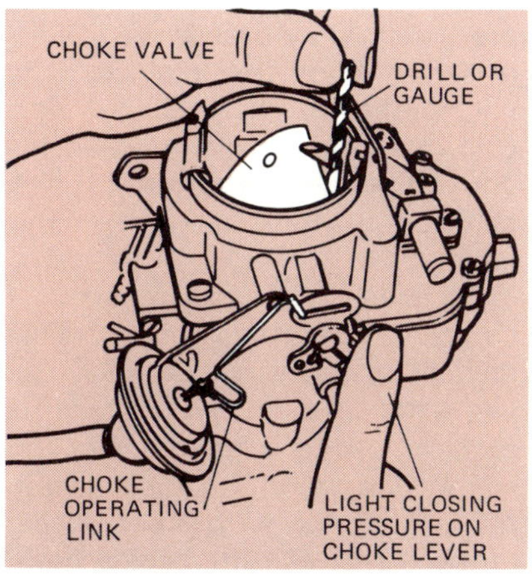


**Adjust choke vacuum break with a drill if you don't have a gauge.**

■ **An air leak** into the intake manifold will lean out the fuel mixture and cause poor acceleration (also rough idling). Squirt a 50-50 mixture of SAE 10W motor oil and kerosene around the base of the carburetor and intake manifold with the engine running at idle. If the mixture is sucked inward and/or engine speed increases, tighten bolts. If this doesn't stop the problem, replace the gasket.

■ **Thermostatically controlled air cleaners** have been on most cars since more than 10 years. A valve in the air cleaner snorkel is controlled by a temperature sensor to provide a leaner fuel mixture and thus lessen emissions. If the valve doesn't work, it may stay closed too long or may not close at all. Hesitation would result.

With the engine cold, look into the snorkel. The valve should be closed. Start the engine. The valve should start opening and be fully open when engine temperature reaches 100° to 110° F.

If the valve isn't working right, see that the hose to the snorkel is tightly attached and not torn. Then replace the temperature sensor.

Noncarburetor-related reasons for hesitation are the following:

**1.** Weak fuel pump.

**2.** Faulty distributor vacuum advance, bad sparkplugs, incorrect ignition timing, damaged distributor breaker points or weak spark.

## Stall out at ground level

An engine can stall under varying circumstances, as follows:

**1.** Right after being started when cold.

**2.** Right after being restarted while hot following a long drive.

**3.** Starts but fails to keep running (cold or hot is immaterial).

**4.** After being started (cold) and while warming up.

**5.** As you accelerate from a standing stop.

**6.** Without warning and suddenly as you are driving.

**7.** In hot, cool or wet weather only.

Once you have classified *your* stalling condition, tracking down the cause is generally easy. In all but a few cases, the cause of the trouble is confined to a carburetor area, another part of the fuel system or the ignition system.

### Troubleshooting stalls

■ **Engine cold:**

**1. Automatic choke.** If your choke system has a vacuum break, check its operation, too. A manually operated choke may also be the cause of a stalling condition. Check the plate for full travel (no binding) and the cable for free movement (no hang-ups).

**2. Fast-idle adjustment.** A fast-idle adjustment below specification is too low to feed an engine the fuel needed to keep running when started in cold weather.

***QUICK FIX:*** With engine warm, hook up a tachometer. Release choke cam so the choke plate closes. Start the engine. Rotate the fast-idle cam so fast-idle screw rests on high step of cam. Engine speed should now be at specified fast-idle speed, about double the engine's specified slow (curb) idle speed.

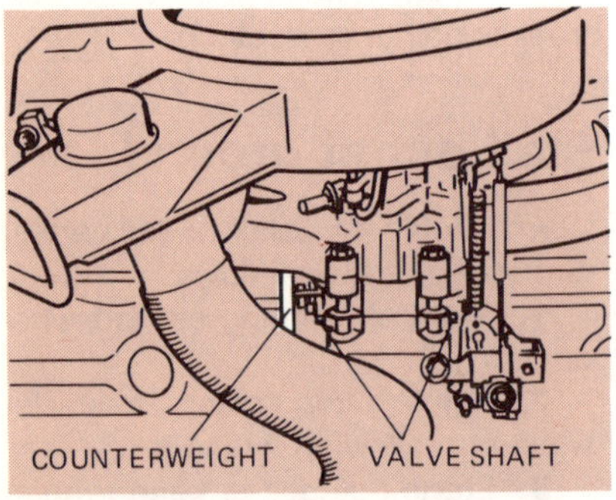


**To find manifold heat-control valve, look for counterweight and two ends of shaft.**

**3. Manifold heat-control valve** (heat riser valve). One stuck open will prolong the time needed for an engine to warm up. This added time under rich fuel mixture conditions can contribute to stalling. Check by trying to move the counterweight, which is normally right beneath the exhaust manifold. *Caution:* The engine *must* be cold to avoid a nasty burn.

***QUICK FIX:*** If the valve is stuck, apply penetrating oil where the valve shaft goes through the manifold and tap (do not bang) the counterweight with a soft-nosed (plastic) hammer. To prevent any recurrence, keep valve shaft lubricated.

■ **Engine hot:**

**1. Slow (curb)-idle speed.** If a hot engine stalls as it idles, the carburetor slow idle speed adjustment may be set too low or the idle mixture may have gone awry, indicating that the carburetor needs cleaning and rebuilding.

**2. Flooding.** The engine may have flooded. You will usually smell gas, especially if you lift the hood. The problem may be caused by a number of conditions. Carburetor percolation is one. The choke plate may be stuck in a partially closed position. Also, the carburetor air-cleaner filter may be dirty (produces over-rich fuel condition); carburetor float may be set too high; carburetor needle valve may be stuck in open position; fuel pump pressure may be excessive.

■ **Added info:** One problem is an engine that starts smoothly when cold, but is hard starting when hot. The engine cranks sluggishly as if the battery or starter is faulty, which they aren't.

Be on the lookout for a buildup of carbon in the engine that gets hot and thereby increases compression, putting extra load on the starter. The problem affects older engines primarily. One way to find out if the trouble is occurring is to make a compression test with the engine hot to see if engine compression is *higher* than specifications.

■ **Starts but fails to keep running:**

**1. Inadequate fuel supply:** Slow-idle speed set too low, an imbalanced idle mixture because of a dirty carburetor air-filter element or dirty carburetor, defective choke operation, restricted fuel filter, improper carburetor float setting, defective fuel pump, carburetor icing or defective fuel pump.

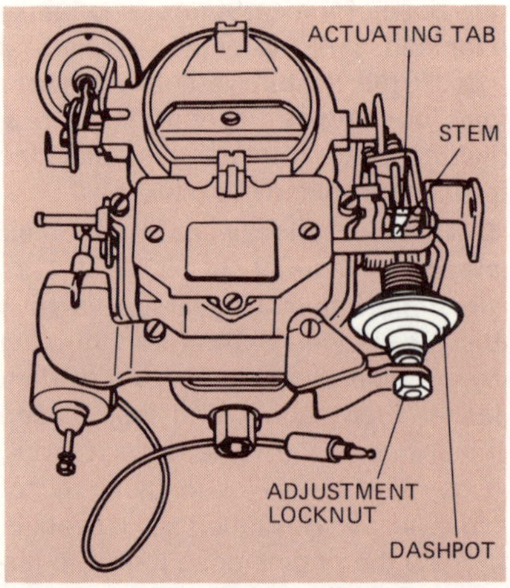


**Throttle return check, or dashpot as it's called, keeps engine from stalling.**

**2. Carburetor dashpot.** On some cars with automatic or manual transmission, the engine has a tendency to stall if the throttle is closed quickly. To prevent this, a dashpot (or throttle return check) is placed on the carburetor to retard the speed of throttle closing. This part consists of a piston or diaphragm and a spring closed check valve. If the dashpot is improperly adjusted or damaged, stalling may occur.

Dashpot stems, which intercept throttle shaft cams by linkage, are adjusted differently depending on the carburetor. If adjustment fails to solve a stalling problem and it is likely that the dashpot is faulty, the part has to be replaced. It cannot be serviced.

**3. Ballast resistor** or primary resistance wire is defective.

**4. Restricted exhaust.**

■ *On acceleration from a standing stop:*

Look for the following carburetor-related conditions, which we have already discussed:

- Clogged fuel filter.
- Defective accelerator pump circuit.
- Low float lever.

Also check for:

- Worn, improperly gapped spark-plugs.
- Defective distributor vacuum advance.
- Grounded distributor primary wire.
- Inoperative thermostat. An inoperative thermostat could delay engine warm-up and possibly contribute to a stalling problem.
- Low engine compression.

■ *On the highway, no warning:*

The way an engine stalls generally pinpoints the trouble area. If the engine stutters and jerks, and then conks out, chances are the problem is a fuel system failure. If the engine just dies suddenly, the trouble area is most likely ignition.

If a fuel system failure is indicated, the problem could be:

- Out of gas.
- Clogged fuel filter.
- Vapor lock.
- Fuel pump failure.
- Dirt (or ice in winter) has wedged between needle valve and seat, causing the engine to flood.

***QUICK FIX:*** You may be able to clear the obstruction without taking off the carburetor bowl cover by tapping the bowl with the handle of a screwdriver in the area of the needle valve.

If an ignition system failure is indicated, look for the problem to be one of the following.

- Ignition system drown-out (wet weather only).
- Ignition wire has come loose.
- Poor vehicle ground.
- Loose electrical connection at distributor tower or coil, loose coil-to-distributor primary wire, or distributor primary wire has become bare and is grounding out against distributor housing.
- Bad coil.

### Backing down backfire

Backfire is defined as a subdued explosion in the exhaust manifold.

From a carburetor standpoint, the thing to do is make sure the automatic choke isn't adjusted too lean and that the car isn't running on gas which is contaminated. Water in fuel leans out a mixture, causing backfiring.

***QUICK CHECK:*** Disconnect the fuel line and allow some gas to fall into a receptacle. Rub some between the fingers, feeling for grit. Let some evaporate. If drops remain, that's water.

If a fuel system has taken on a load of contaminated gas, drain and flush the system. Install a new filter.

Other causes of backfire include intake valve trouble and crossed sparkplug wires.

Although generally regarded as backfiring, a puttering sound from the tailpipe is technically called "afterburning." This condition is usually caused by leaking exhaust valves.

## OTHER PROBLEMS

Here is a summary of what usually causes some other carburetor-related problems:

**1. Dieseling.** This term describes an engine that continues to run after the ignition is shut off. Troubleshoot by checking the following:

- Idle-stop solenoid on many (not all) carburetors. Watch the solenoid plunger as someone starts the engine. It should rest against the throttle lever. Turn off the ignition. The plunger should retract at once. If not, replace the part.
- Dirty, binding throttle linkage.
- Improperly adjusted slow-idle speed.
- Manifold heat-control valve is stuck in the closed position.
- Improper distributor point gap and dwell, and ignition timing.
- Engine is running hot.
- Buildup of carbon in engine.

**2. Rough idling.** Consider the following if your car reminds you of a ride on a New York subway train:

- Dirty carburetor air-cleaner element.
- Incorrect slow-idle adjustment.
- Improper idle mixture.
- Stuck automatic choke.
- Improperly set float.
- Partially clogged fuel filter.
- Dirty carburetor.
- Stuck manifold heat-control valve.
- Engine vacuum leak.
- Faulty distributor breaker-point gap and dwell angle.
- Plugged positive crankcase ventilation valve.

***QUICK CHECK:*** Remove the PCV valve from its seat and shake it. You should hear it click. If not, replace it.

**3. Misfiring.** The cause of misfire usually is to be found in the ignition system, yet sometimes a carburetor-related condition has a bearing.

Consider an improper fuel mixture, dirty carburetor air cleaner and closed choke. Other problems previously discussed (other than ignition) that cause misfiring are an air leak at the intake manifold, stuck manifold heat-control valve and PCV system restriction.

### Weather-related trouble

Some owners have a problem with weather affecting carburetors. Two conditions prevail: percolation and carburetor icing.

**1. Percolation** refers to gasoline in the carburetor bowl boiling over into the intake manifold. Excessive heat bearing on the carburetor, usually after a long period of operation in hot weather, is the cause. Percolation is often called vapor lock, but the two conditions are different.

Vapor lock is the formation of vaporized fuel pockets (bubbles) in the fuel system caused by heat. Bubbles block the flow of fuel to the engine, causing the engine to stall. Vapor lock incidents in modern cars are rare.

Percolation also is no longer widespread because of the inclusion on many carburetors of relief valves. However, it can occur if the relief port plugs.

After a hot engine is shut off, percolation may develop. Attempts made to restart the engine result in hard starting. Allow the engine to cool a bit. Then, holding the accelerator to the floor so the throttle is wide open, crank the engine. It should start.

***QUICK FIX:*** Check the antipercolation valve for clogging and to see that the valve cover isn't stuck.

**2. Carburetor icing** is the formation of ice in the carburetor that cuts off air and causes the engine to flood and stall. The condition generally occurs when the ambient temperature is between 30° and 50° F. and the relative humidity is 65 percent or more. If your carburetor is prone to icing, add a commercial anti-icing additive to the fuel.

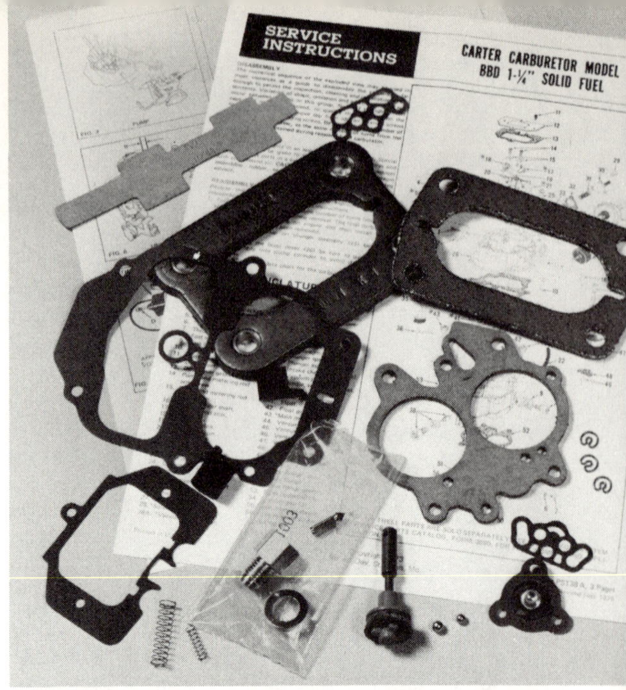


**TYPICAL TUNE-UP** kit has gaskets, accelerator-pump plunger, needle-seat assembly, check balls for intake and discharge valves, diagrams, instructions.

# How to zip-kit your carburetor

■ WITH AN ORDINARY amount of mechanical ability, you can take your carburetor apart and recondition it.

Two or three times in its life a carburetor needs more than adjustment—it needs to be torn down and rehabilitated with a tune-up kit, a so-called zip kit that includes gaskets, valve balls and other parts that wear or deteriorate.

To give you some initial confidence, remember that there is just one adjustment to be made inside the carburetor: setting the float height. There are only six or seven more adjustments after the carburetor is reassembled.

There are two general rules: The first is that cleanliness inside a carburetor is crucial to its operation.

The second rule is to follow precisely the instructions that come with each tune-up kit.

The exact sequence for reassembly and external adjustment is important. Carburetors are made in such a way that you can't skip one step and then come back to it later on.

The first thing you should do, because there is a chance of gas spillage in the engine area, is to ground the distributor-to-coil wire to prevent sparks.

After you do this, purge the disconnected fuel line by flushing the gas into a suitable container; just turn the engine over a few times with the starter. This removes dirt that is loosened when the fuel line is disconnected. After the carburetor is removed, cover the manifold inlet to keep dirt out of the engine.

Throttle plates at the bottom of the carburetor are seated with precision. Therefore, the carburetor must be raised above the workbench so that if the throttle mechanism is moved, the throttle valve will not be pushed against the bench and become damaged. If you don't have a carburetor stand, one can be improvised with bolts and nuts as legs.

You should be prepared with a cleaning solvent. For heavily used carburetors with burned-on carbon, a strong automotive parts cleaner and a firm brush are necessary. Rubber, leather or plastic parts should not be put in solvents or cleaners.

Before removing the carburetor, be sure you can identify all hoses connected to it. If necessary, label them or draw a quick picture.

Remove all external linkages. Next, remove the choke pull-off and its bracket. This device works with the automatic choke to pull it open by vacuum. Its diaphragm can deteriorate, causing poor performance. A simple test is to press the diaphragm in and hold your finger over the vacuum hole. The diaphragm should stay depressed until you remove your finger (see photo). Its full operation can be tested if you have a vacuum pump.

The choke pull-off and the automatic choke are among the first items to check when there is a starting problem. Similarly, when the complaint is, "She doesn't run right until warm," you check the entire choke circuit. The bimetal control in some automatic chokes can fatigue with excess engine heat, the presence of dirt and exposure to the weather.

While taking apart the carburetor, watch for dirt and wear. Under the dust cover you'll find the accelerator pump, metering rods and the countershaft that operates them. These parts

SEE ALSO

**Chokes, auto . . . Dieseling, auto . . .
Differentials, auto . . . Distributors, auto . . .
Emissions controls . . . Engines, auto . . .
Fuel systems, auto . . . Gaskets, auto . . .
Misfiring, auto . . . Stalling, auto . . . Tune-up, auto**

**WHEN YOU BEGIN** to reassemble the carburetor, carefully insert the pump plunger into the pump cylinder as the bowl cover is replaced.

eventually wear, but do so more slowly than the items in the kit.

Next, you remove the fuel-bowl cover, actually the entire top casting. In the fuel bowl are the floats, pump plunger and, among other things, two small valve-check balls that can be replaced from the kit.

Test the floats to make sure no fuel has leaked into them. Float leaks cannot be repaired; new floats are necessary if there is a leak.

Opposite the fuel bowl are the venturi clusters. The two screws that hold the venturi have fuel passages through them that are precision-calibrated. These should be blown clean. Use a compressed air source.

When you get to the bottom or throttle-valve casting, you will find the two idle-adjusting screws. Pry off the plastic limiter caps. If new ones are needed, replace them from the kit. It is important to establish the setting of these screws before removing them. You do this simply by screwing them in until they just seat, counting the turns on each. When you replace them, screw them all the way in and then back out the same number of turns at which they were set before.

The caps are a legal and protective feature that prevent adjustment to too-rich or too-lean a mixture. Follow directions to install new ones.

After soaking parts, dry and blow them clean. Each jet, tube and passageway should be blown out.

All gaskets should be replaced with those from the tune-up kit. Fresh, soft gaskets are a must to insure a perfect seal. Be sure the mating surfaces of the castings are clean and undamaged.

In the reassembly phase, the first adjustment is float level. A gauge and chart for proper float level are in the kit (see photo). Put in the floats and hold them in place with the U-shaped retainer. Be sure you hold this retainer against the bottom of its slot while gauging. Hold the floats in place by lightly pressing the float tang against the seated needle at the fuel entry hole. To adjust, bend the tang on the float arm.

When you come to replace the bowl cover, care must be taken with the assemblies on the pump countershaft. Before putting the cover on the bowl, insert the spring on the stem of the new pump plunger. Then insert the plunger stem through the hole in the bowl cover on the fuel bowl. The pump plunger must be guided carefully into its cylinder so that the leather is not damaged.

Guide the vacuum piston with its attached metering rod arm and then the pump arm.

With dust cover screwed on, bolt carb to manifold, replace linkages and hoses, and tune idle and mixture per kit instructions.

**FLOAT MUST** be set to specification using the gauge included in the kit. The float operates the needle valve that controls fuel flow to the carburetor. Bad adjustment can cause flooding or fuel starvation.

**CHOKE PULL-OFF** can be tested by depressing diaphragm and placing finger over vacuum hole. A good diaphragm will remain depressed for as long as your finger is kept on the hole.

**A GAILY COLORED** stripe pattern is available in both warm and cool colors as a covering for your patio or wood deck.

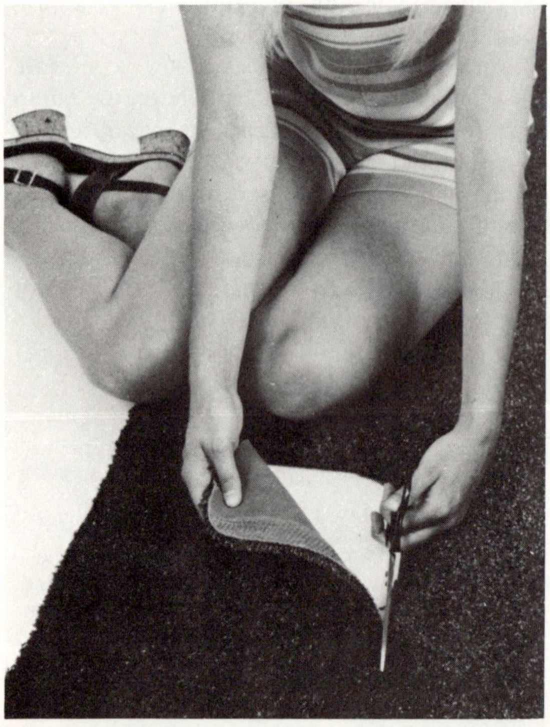

**THE CARPETING** can be cut with ordinary scissors and does not require stitching where adjacent pieces meet.

■ IF YOU WANT to change the appearance of your back yard but would like to complete the project reasonably quickly so that you can enjoy it without having to wait, consider the idea of using carpeting that's designed specifically for outdoor use.

The carpeting is available in solid colors and in striped, multicolored patterns, as well as in a realistic-looking grasslike texture called Lawnscape, shown above in the yard of former baseball star Mickey Mantle. (You'll have to admit that having a carpeted yard beats lawnmowing!)

By covering your outdoor "floor" with carpeting, you can make a patio or deck as handsome as any room inside your house. Outdoor carpeting is easy to work with; in fact, it doesn't cost you anything to install it, because you can do the job yourself. The material can be cut with a pair of scissors.

Among the advantages of outdoor carpeting is its versatility. It can be laid over any existing wood deck, concrete patio or other rigid surface. Adjoining strips can be butted together without requiring any stitching. If you find it necessary to keep the edges of the material from curling, a double-sided carpet tape can be used for that purpose.

One word of caution: Lawnscape, the synthetic grasslike material, should be laid over an asphalt or concrete base. You'll find that it covers a surface quickly—it comes in 6 and 12-foot widths.

Any of the outdoor carpetings mentioned on these pages can be installed permanently with special adhesives supplied by the manufacturer.

**MICKEY MANTLE,** former star of the New York Yankees, chose Lawnscape as a poolside covering. He uses the patio to keep his putting eye sharp.

**It wouldn't take long to cover your outdoor "living room" with a bright new face. Best of all, you can do the work yourself.**

**By HARRY WICKS**

# Spruce up your patio with indoor-outdoor carpeting

**IN THIS HOUSE,** the carpeting runs from the family room out onto the patio. The result is a continuity of the decorating theme indoors and out.

**THIS SUNKEN** concrete patio was updated by adding bright colors. The use of carpeting with bold strips makes the patio appear larger than it is.

SEE ALSO

**Barbecues . . . Benches, deck . . . Cabanas . . . Cookout bars . . . Decks . . . Firepits . . . Garden shelters . . . Gazebos . . . Grills . . . Patios . . . Poolhouses . . . Swimming pools**

# Car-top boat booster

By MARK SOSIN

■ A CARTOP BOAT BOOSTER for under 10 bucks is a bargain even if you don't have a bad back like fisherman Les Hodax. To handle his hull without help, he designed the simple system shown.

Most boats are slightly wider than standard rooftop carriers. You can mount 2x4s to extend your carriers a few inches more than the distance between the gunwales of your boat. Mount the 2x4s to your carriers with through-bolts, but countersink the bolt heads.

In addition, you will need two lengths of light-walled pipe about one inch in diameter and two bolts from six to eight inches long. The length of the pipes will depend on the height of your vehicle. They should slant to the ground without touching the side of your car. One end of each 2x4 is notched out to receive the pipe in the center, and that end is rounded off to make sliding the boat on and off easier. A 1/4-inch hole is drilled through this end of each 2x4 and also through one end of each pipe. Now all you have to do is lay the pipe in the notch, slide the bolt through the matching holes, and your pipe rails are ready for the boat.

To prevent the boat from sliding too far, and to help secure it while traveling, Hodax loads from the right side and has mounted angle irons on the left side of his 2x4s to hold the boat in place.

SEE ALSO

**Boat camping . . . Boating, trailers . . . Campers . . . Camping . . . Canoes . . . Kayaks . . . Pontoon boats . . . Sailboats**

**CARTOPPING A BOAT** takes a lot less brawn when slide pipes are fastened to the roof brackets.

**THE HULL CLEARS** the side of the car, allowing the gunwales to ride at an easy angle onto the carrier.

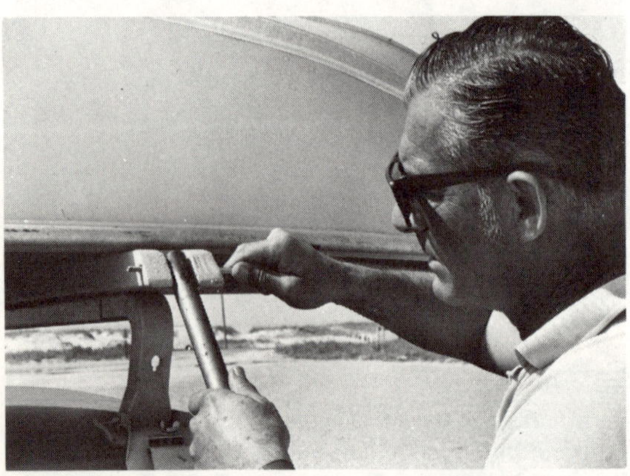

**THE ROOF RACK** is drilled on the right side of the car to take a bolt securing the pipe.

**ANGLE IRONS** on the left side of the roof rack prevent the boat from sliding too far.

**A METAL CATCH** is bent to fit the hood or trunk lip. An opening is cut and bent in the catch.

**THE TIE-DOWN ROPE** is measured for a tight fit and permanently knotted to a cutout in the clip.

**SECURE** without a spiderweb of tie-down lines, this cartopper is held in place with ropes to clips under the hood and trunk-lid edges.

By MARK SOSIN

# Easy tie-downs for a boat

■ TIE-DOWNS, too often, are a tangle of ropes that have been woven in a spider web over and around a cartop boat or other oversize gear on the roof. Then they come down and through windows and around door handles and bumpers in an attempt to keep the boat or luggage from shifting forward and aft or side to side while traveling down the road. Sometimes the rigging is so extensive that doors, hood or trunk of the car can't be opened until the launching ramp is reached. Checking the oil, unloading a spare tire or even getting out for a stretch presents a problem.

But Les Hodax of New Shrewsbury, N.J., worked out this simple solution. He fashioned light-gauge metal clips, as shown, that seat in the joint between car hood and fenders and in the recesses at either side of the trunk lid. A rectangular opening is cut in each catch to hold a premeasured line knotted to it. Position catches, close hood and trunk and the boat is secured. Open them for instant release.

**ENGINE SERVICING** is a cinch since neither metal catches nor tie lines are removed when raising the hood.

SEE ALSO

**Boat camping . . . Boating, trailers . . . Campers . . . Camping . . . Canoes . . . Kayaks . . . Pontoon boats . . . Sailboats**

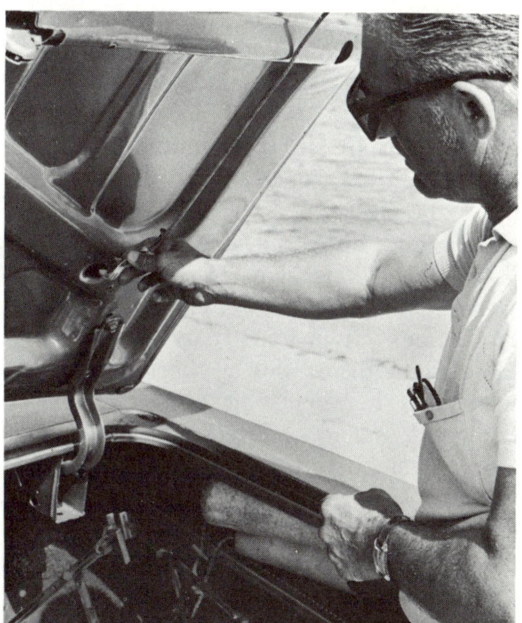

**TRUNK ACCESS** is even more convenient. With no tie-downs knotted to the trunk handle or rear bumper, gear can be unloaded at any time without unlashing the cartopper. Tension on the transom remains the same each time the trunk lid is closed.

**FILM LOADERS** and accessories include (clockwise from upper left): Premier, Prinz 66, Western, Watson, unbranded model, Prinz standard. In the front row are empty cartridges, bulk film and the new Whitehead Cinek self-loading spools still in pilot production stage

# Load your own 35-mm film cartridges

**By IVAN BERGER**

■ WHAT'S THE BEST photo equipment investment you can make? If you use a 35-mm camera, the answer is a bulk loader. Fill it up, use its contents once, and it has paid for itself. It will keep on cutting your film costs a third or more. And you'll be able to use some special-purpose films you just can't buy by the roll.

Here's how the economics work out: With a loader costing between $6 and $20, plus 18 empty cartridges (less than a quarter each—and they're reusable), you can turn a 100-foot roll of film into 18, 36-exposure cartridges. An under-$50 bulk roll of Ektachrome, for instance, gives you 18 rolls at less than $3 each, including cartridges, a saving of well over a dollar a roll, or about $20 total.

Exact costs vary. Discounts are available on both factory loads and bulk rolls. The number of uses you can get from a cartridge depends on how carefully you (or the lab you send your film to) open them, and how free you keep them from dust. You can even reuse factory-loaded cartridges—except for Kodak's, which must be pried, not popped open, and can't be reused.

You can bulk-load quite a number of the films you now buy in cartridges—plus several that don't come in factory loads. There are black-and-white films from Kodak, Ilford, Luminos and Supreme, plus the super-sharp, almost-grainless H&W Control films. In color, you can get Kodak Ektachromes and Vericolor II (but not Kodachrome or Kodacolor).

Most well-stocked photo stores carry at least some bulk films (usually Kodak's), as do such mail-order shops as Spiratone (135-06 Northern Blvd., Flushing, NY 11354), Porter's (Box 628, Cedar Falls, IA 50613) and 47th-Street Photo (36 East 19th St., New York, NY 10003).

Bulk rolls come in lengths of 100, 50 or 27½ feet. The large, economy-size 100-footers are really the most economical, but you may prefer shorter rolls of films you use less often.

Today's loaders are light-tight and require darkness only for the few moments it takes for you to put the bulk roll in. If you lack a darkroom, you can use a light-tight changing bag; in a pinch, you can even do it at night under the covers of your bed or in a closet.

Actually, there are two light-tight compartments in these loaders: The large one holds the bulk roll, and the smaller, with its own door, holds the cartridge. Connecting them is a light-tight slit, the light trap, designed to let the film be drawn out of the big compartment.

To load a cartridge, you first open it and remove its spool, then attach the spool to the film end protruding through the light trap. With the film coming from your right, the spool's long end should point toward you.

Reassemble the cartridge around the spool, insert it under the crank in the loader, close the light-tight door, and wind till you have enough film in the cartridge. Then open the door, remove the cartridge, and cut and trim the film end. Some cameras will require that you trim your leader like those on factory loads or 27½-foot rolls, and you can get trimming templates for this. Most cameras will do nicely with a short, diagonal cut as shown below.

There are two basic types of loaders: Inexpensive, under-$10 ones, such as the small Prinz (and the unbranded model shown) use felt-lined slots as light traps. The felt works, but the longer you own the loader, and the more you use it, the greater the chance that a sharp bit of dust will embed itself in the felt and scratch your film.

More expensive models (still under $20), such as the Watson, Western, Premier and Prinz 66, have light traps that can't scratch the film because they never touch it. Instead, a door clamps tight against the film when you open the cartridge compartment, but opens wide to pass the film through freely when the cartridge-compartment door is closed. Interlocks prevent opening the cartridge compartment till you've closed the light trap; but you must remember to open the trap when you wind the film to prevent scratches.

These models also have counters that click off the number of exposures you've wound; felt-trap models usually just have charts telling how many crank winds correspond to what number of exposures.

Most models of the same type work about equally well, though the tiny, light unbranded model would chew film if you set its lid at the wrong angle, and the Premier had the best interlock, but the worst crank.

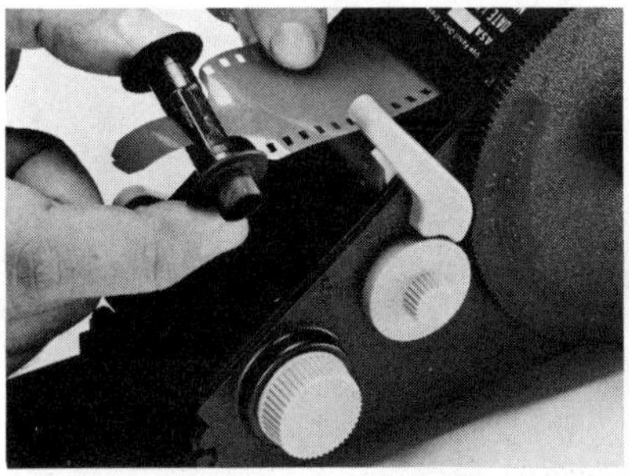
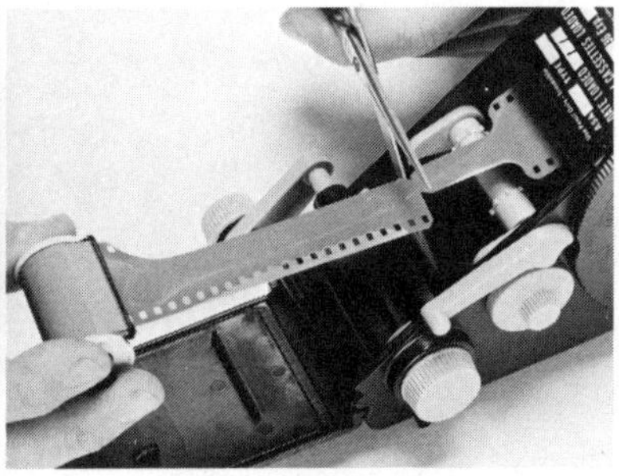

**SHORT** 27½-foot rolls have a precut tongue to fit slotted cartridge spools (left), a film leader and a new tongue every 36 exposures. Longer rolls save you more

**LONGER,** 100-foot and 50-foot rolls must be taped to the film spool (left) and cut at the end of each roll (center). The leader must also be cut in the end (right)

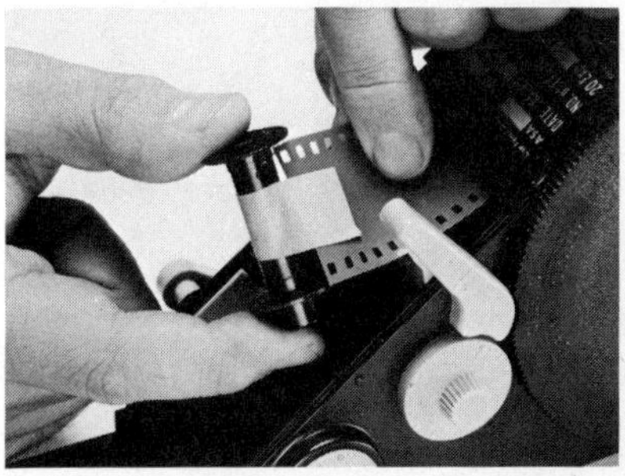
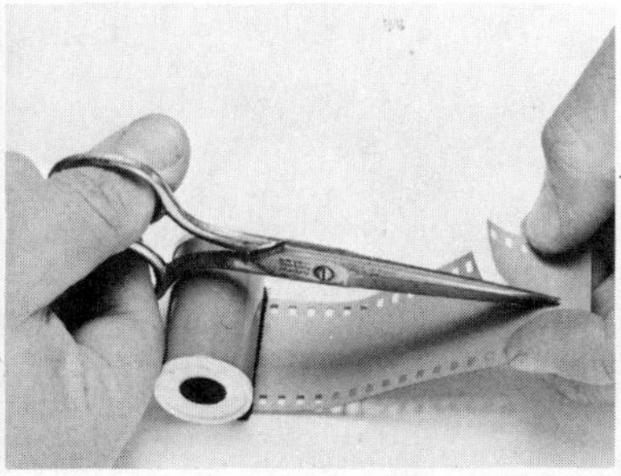
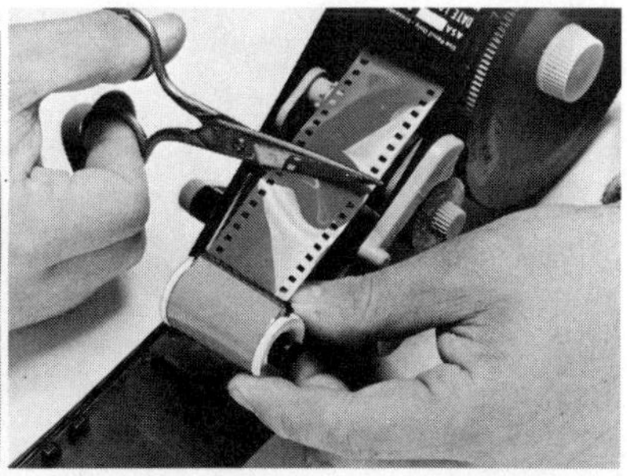

# Recorder repairs you can make

**Cassette tape recorders are not as complicated as they seem. Much of the preventive maintenance and many of the basic repairs you can do yourself. All you need are the proper tools and know-how to get started**

By FOREST BELT

SEE ALSO

**Cassette tapes . . . Citizens band radio . . . Hi-fi centers . . . Mixers, audio . . . Speakers . . . Storage ideas**

CASSETTE RECORDERS may seem complicated, but actually they're basically simple. They all work pretty much alike, and in most cases their parts are readily accessible. Before you throw up your hands at a balky machine, try these quick and easy fixits. Chances are, they'll get an ailing machine back in shape without your having to resort to major repairs.

Weak or dead batteries in a portable recorder

**HERE'S A VIEW** of a typical cassette recorder out of its case and shown from the bottom. Your recorder (and those shown on the following pages) may not resemble this model, but you can quickly identify the mechanical parts and their functions. In general, these are the parts you can repair.

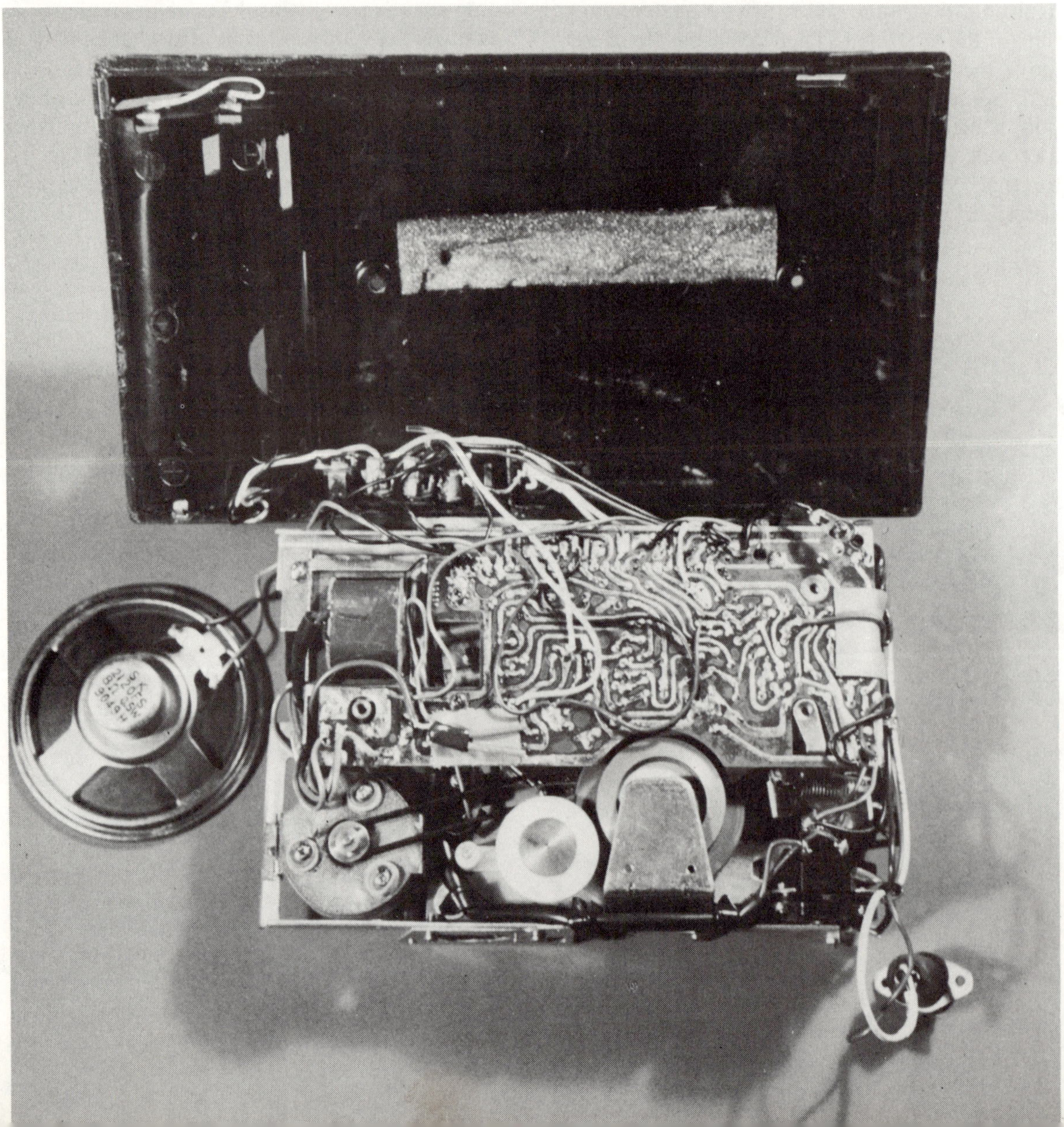

**JUST DROP IT IN** and the special cassette tape above automatically wipes the heads clean as it runs through the machine. Such aids, while not a substitute for more thorough hand cleaning, are convenient for occasional use. Also available are test tapes that check for wow and frequency response.

cause more problems than you may realize. This may sound a little silly—like checking to see if the power cord is plugged in on a dead appliance—but it's such a simple thing it's often overlooked—just like the power cord that *did* unknowingly pull loose. Run-down batteries cause variations in tape speed that produce "wow" sounds or make the musicians seem to be playing in a tub of molasses.

If the drive spindles turn normally when the machine is empty, but stop or slow down the minute you insert a cassette, this generally indicates the batteries are weak—just strong enough to drive the spindles freely, but unable to take the added load of the tape.

If your recorder has a 117-volt a.c. adapter, make a quick check by switching to this. If it performs well on house current, the batteries are obviously the culprit. In replacing them, change the whole set—batteries in series wear out about evenly. Also make sure the battery terminals are clean and make firm contact.

When the machine won't budge with a tape inserted, but purrs along smoothly when you take it out, the cassette itself may be at fault. The tiny tape inside sometimes snarls or spills so it runs sluggishly or jams the drive mechanism. The answer here is to try several other cassettes. If the machine checks out okay on these, you can blame the first cassette. But don't throw it away. A snarled tape can often be cleared by running it rapidly back and forth. Try to get it to move on "Fast Forward," then reverse it and run it backward on "Rewind." Do this several times, then see if it will run through at normal forward speed. If your recorder doesn't have a Fast Forward, rewind the cassette in one direction, then flip it over and rewind it the other way. This has

**WEAK BATTERIES** can cause more problems than you think. Replace them frequently and in sets—not just one at a time. Be sure the cells are correctly inserted according to the polarity markings inside the case.

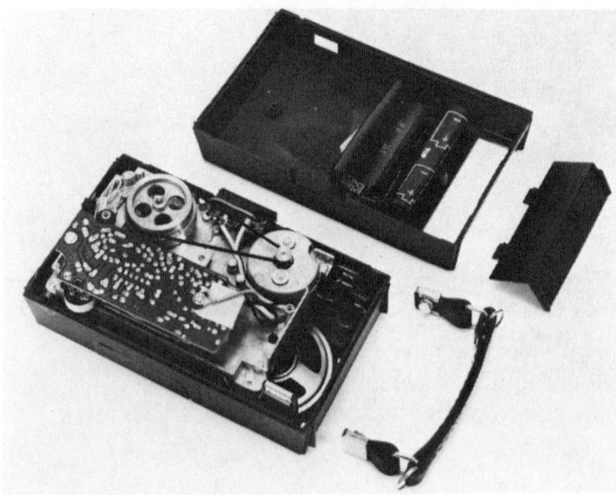

**REMOVING THE BOTTOM** of the case usually exposes the underside of the chassis, motor and flywheel (above). It may also release the handle and other small parts. Note how they fit together to facilitate reassembly. Using a cotton swab dipped in alcohol, clean oil and grime off the motor pulley and flywheel (below). Such dirt can cause belt slippage, varying the drive speed.

the same effect as running the tape forward at fast speed.

The next most common cassette troubles are the periodic buildup of magnetism on the heads and the dirt and grime that collect on the heads and in the drive mechanism. Dirt on the heads prevents the tape from making good contact, clogs the delicate head gaps and quickly reduces sound quality. Oily grime on the capstan, pulleys and drive belts causes tape slippage that, like weak batteries, results in wow, erratic tape speed and annoying fluctuations in musical pitch.

Cleaning and demagnetizing the heads can be done on most recorders without dismantling them. This should be done at least twice a year—more if you give your machine heavy use. The heads pick up magnetic charges as the tape moves past which interfere with proper recording and playback of magnetic signals on the tape, resulting in poor sound reproduction.

Small demagnetizing tools cost a few dollars at electronics and hi-fi shops. You plug the tool into 117-volt a.c. and move its tip back and forth a few times across each head. Keep the tool from actually touching the heads to avoid damaging the delicate surfaces. The tip should just brush lightly by them. Be sure no prerecorded tapes are lying nearby as they may be demagnetized and erased accidentally.

For cleaning the heads, you can use ordinary denatured alcohol or special solvent sold for the job. Apply it gently with a cotton swab, as shown in the accompanying photos. Particles of mag-

**TO REACH THE TOP SIDE** of the mechanism, remove the screws holding the chassis frame to the case, then carefully lift the chassis out and turn it over. Watch out for fine wires that may run to battery contacts in the case.

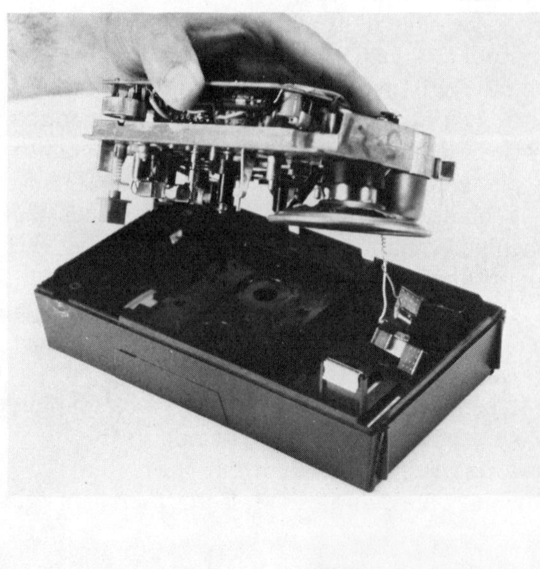

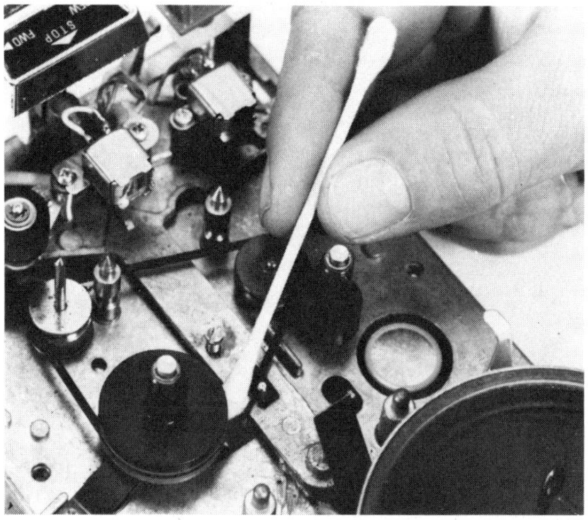

**CLEAN UPPER PULLEYS** in the same way as the lower ones with an alcohol swab. Also wipe the capstan, pinch roller and tape guides. Gently check the belt tension.

**HEADS CAN BE CLEANED** with a special solvent or ordinary denatured alcohol. Wipe them alternately with wet and dry swabs until no traces of dirt remain.

netic oxide rub off the tape during playing and must be thoroughly cleaned out of the tiny head gaps or they'll clog them. A silicone head lubricant is also available to provide a smooth, gliding surface for the tape. Other aids you can buy include test tapes for checking on various performance characteristics and head-cleaning tapes that do the cleaning job for you. The latter are handy, but should not replace an occasional thorough hand cleaning.

For inside cleaning, you'll need to remove the mechanism from its housing. In some models, the chassis is exposed by removing a cover. In others, the chassis lifts out of the enclosure from the top. Be careful on the more intricate autoreversing and changer models not to damage the delicate machinery. Portable machines are much simpler. Their housings are usually in two shell-like halves. You remove the bottom shell first, exposing the underside of the chassis. Then you take the chassis out of the top shell and turn it over to reach parts on the upper side. Battery contacts are usually in the form of springy clips that can be slipped off plastic posts to free the chassis.

Using swabs and solvent, clean all pulleys, capstans, pinch rollers, spindle hubs, drive belts and other moving parts. Gently check the belts for good tension. Replace any that are loose. Exact replacements can be ordered through hi-fi dealers or the maker of your particular recorder. But don't attempt to open a machine still in warranty as this will void it. Problems involving extensive repairs are jobs for a trained serviceman.

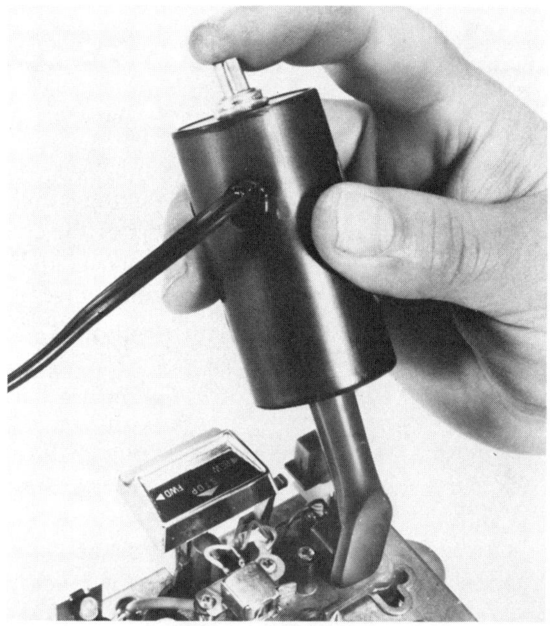

**PERIODIC DEMAGNETIZING** of the heads is easy with a small demagnetizing tool like Audiotex's above. The tip should just brush past—but not actually touch—each head. The pinch roller should press the tape firmly against the capstan for a smooth, uniform drive. You can check the pressure by drawing a thin strip of paper between the roller and the capstan, as shown below.

**TOOLS FOR** cassette repair include: Audio Accessories kit (far left) with a spare shell, splicer and splicing tabs; 3M kit (top left) shown in more detail below; Editall splicing block (left) with splicing tabs, a special tape-marking crayon, and razor blade. The 3M kit (photos below) includes strips that fish broken tape ends out of the cassette shell, a splicing block to repair broken ends, and a winder to take up slack in the tape after the completion of repairs.

# Snarled cassette tapes can be fixed

**By ROBERT CHAROLAIS and IVAN BERGER**

CASSETTE TAPES don't break or snarl as often as they used to—improvements in cassette construction have seen to that. But the cheaper cassettes still are often troublesome, and even the better ones cause occasional problems—invariably on recordings that simply can't be duplicated.

The most common cassette problem is the jammed cassette whose tape won't move. The cause usually is friction inside the cassette, and an uneven "pack" of tape around the cassette's internal hubs. But the cure is usually simple: Hold the cassette three or four feet above a hard surface (such as a desktop or an uncarpeted floor), and drop the cassette so that it lands flat. Do this several times, including at least once on each side. It won't damage the cassette, and will frequently cure the jam-up.

If it doesn't, you'll have to open the cassette. If it's held together with screws (usually five small Phillips-head ones), you're in luck. Place the cassette flat and screw-side up on a table with plenty of clear working space. Carefully remove the screws, the top shell and the plastic slip-sheet inside. If your luck still holds, and the tape hasn't broken or stretched, you can simply smooth it out and wind it back in place.

There seems to be no good way to open welded cassettes; the best we've found is to stand the cassette up on its broad edge (the one through which the tape shows), rest a screwdriver on the seam in one corner (between the

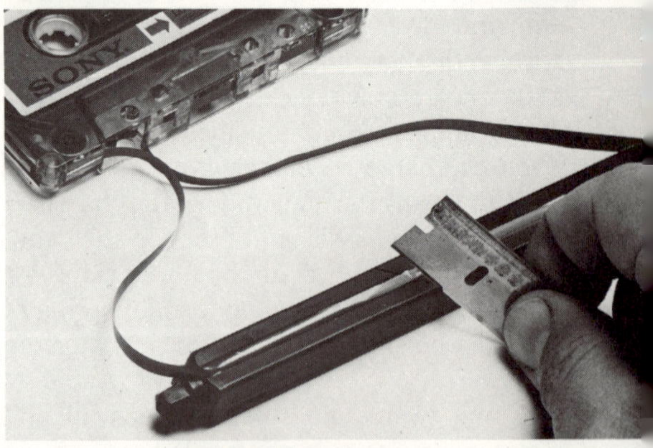

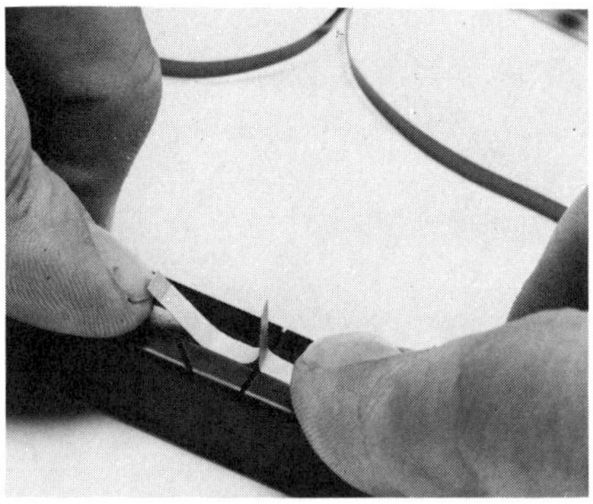

**SPLICING TABS** to make repairs are easier to apply than splicing the damaged tape on the rolls. Splicing tabs need less trimming and pick up less finger oil.

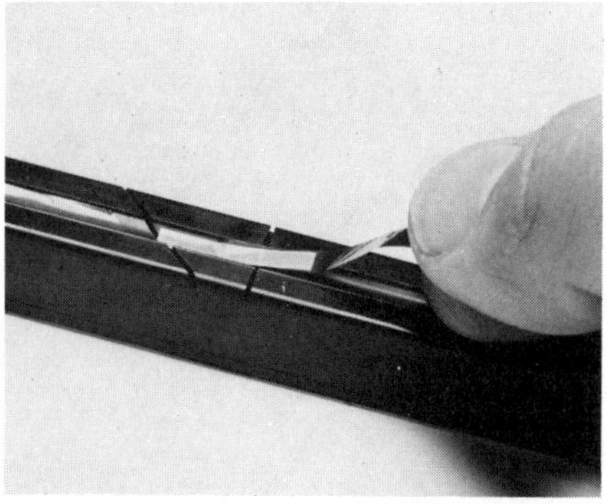

**THE SPLICING BLOCK** holds the tape in a precision channel while you cut and trim with a razor blade. With this type of block it's easy to flick out the tape.

tab pushout and the right edge), hammer the screwdriver till the seam opens, then force the screwdriver down the edge to break the rest of the seam. Repeat this on the other side and, if need be, along the back seam.

**tool no. 1: the spare-parts kit**

If you've just had to blast open a welded cassette, or if parts of your cassette are broken or missing, or if the problem is due to such corner-cutting as using fixed plastic pins instead of tiny rollers to guide the tape, you'll need a spare-parts kit like the Maloney Audio Accessories kit shown, or the similar ones made by Robins and Certron.

These kits are actually spare cassettes with everything but tape in them. There are two shells, two plastic hubs, the slip-sheets which prevent friction and static electricity from building up inside the cassette, tiny teflon rollers and stainless-steel pins to hold them in place, pressure-pad assemblies, and metal shields—plus, usually, a cassette splicer and splicing tape.

As needed, you can replace missing or defective parts from one of these kits, or lift the entire tape pack and takeup hub from the old shell to the new one. If that's the case, thread the tape carefully around the rollers, inside the plastic tape guides and in front of the pressure pad behind the tape window at the cassette's edge.

**tool no. 2: the 3M repair kit**

If the only problem is broken tape, you don't have to open the cassette up—not since 3M brought out its handy Cassette Editing and Repair Kit. The kit includes pliable plastic wands with sticky ends with which you can fish down into the cassette shell to pick up the tape end—usually after several tries—and bring it out for splicing (more on that shortly). Once the repair's complete, you fit one end of the kit's case into one of the tape hubs to wind up the slack tape you've pulled out for splicing.

**tool no. 3: the splicing block**

Splicing any tape requires a device to cut broken tape ends neatly and to hold those ends in alignment while the splicing tape is applied and trimmed. There are lots of splicers around for cassette tape (regular-tape splicers can't be used—cassette tape's too narrow), some with two arms that hold the tape while a third arm comes down to trim the tape ends or the splicing tape, others are simple blocks with channels that hold the tape while you cut and trim with razor blades. Most professional tape editors use blocks like the Editall, and I found it the easiest to use of the three splicers shown on page 648. The Editall's weight keeps it steady (you can also screw it down), its slot is deeply undercut to hold the tape securely, its aluminum construction keeps razor blades from widening the cutting slits and its splicing tape tabs are very easy to apply.

The 3M kit's splicing block, molded into the kit case, held the tape as securely, but didn't sit as steadily or resist cutting like the Editall.

When splicing, always be sure the tape's shiniest side is up, and that neither end is twisted. Then overlap the ends by half an inch or so, make a diagonal cut with a razor blade (or your cutter's built-in splicer, if it has one), remove the loose ends, butt the clean-cut ends together, and attach the tape. Trim away any bits of splicing tape which overlap the edges of the tape; overhead-arm splicers do this for you, but with blocks, you have to run your razor blade along the undercut lip of the tape-holding slot to do it.

**MASTER CARVER,** Tony Walder of Record Ridgway Tools, carves a bowl. The four finished projects explained on these pages are shown on display on the workbench in front.

# Four projects for wood carvers

**By PENELOPE ANGELL**

■ CARVING AND FINISHING even a simple object is one of the most satisfying forms of woodworking.

Many carvings are made with just two basic types of tools: gouges and chisels. Gouges are made of steel that is curved to varying degrees to "gouge" out excess wood. A "quick" gouge has a very deep curve that's intended especially for use in deep and rough cutting. "Slow" gouges have flatter curves and are used in finishing work. Widths at the cutting edge of these tools can vary from 1/16 to 5/8 inch.

Chisels are beveled on both sides and used by carvers to cut in designs and clean out wood in detailed areas during the finishing stages. Their ends can be square or skew (angled) in widths to accommodate rough or fine work.

Your first purchase of carving tools should include three or four sizes of gouges and chisels, a mallet to drive the tools, clamps or carver's screws to secure your work, and sharpening stones.

SEE ALSO
**Birds, carving . . . Candle stands . . .
Hobby centers . . . Internal carving . . .
Jewelry boxes . . . Laminating wood . . .
Marquetry . . . Miniature houses . . .
Modelmaking . . . Veneering . . .
Weekend projects . . . Wood sculpture**

# 1 SNACK DISH

Scooping out this handy dish is a good way to become familiar with your carving tools and how to handle them. You'll need a 9 x 6 x 1¼-in. block of wood. Mahogany or walnut both have fairly even grain and finish nicely. Very sharp tools will make your carving easier and safer.

**Tools:** 1-in. fantail gouge, ⅝-in. bent gouge, ⅝-in. skew chisel, wood mallet and bandsaw or other saw. We used tools from Marples Blue Chip Sculpture set of Record Ridgway Tools.

**Method:** Draw the dish design either directly on the wood or on paper and transfer it. This design is a freehand "avocado" shape with part of the rim slightly thicker to serve as a handle. *Note: In this and the following projects the wood should be securely fastened to a worktable so you can work with both hands safely behind the cutting edge of the tool.*

With the fantail gouge (the working end of this gouge is fanned out) and wood mallet, hollow out the dish. Then with the bent gouge finish hollowing and giving texture to the inside of the dish. (This gouge is actually bent and was designed to hollow shallow areas.)

Next, saw or use a gouge to remove excess wood. Work so that the sides are cupped inward toward the bottom. Then turn the dish over, secure it, and with the gouge shape and give the outside texture.

Last, the finishing decoration, a series of triangles, is carved on the dish top. Draw the number of triangles you wish as in the illustration. Then cut along the center lines with a skew chisel held vertically, tapering the cut from ⅛ in. at the center (B) to nothing at the triangle tips. Cut A to B, B to C and B to D. Remove the waste, tapering and paring the wood away from the sides to the center. Pare A to C, A to D and C to D.

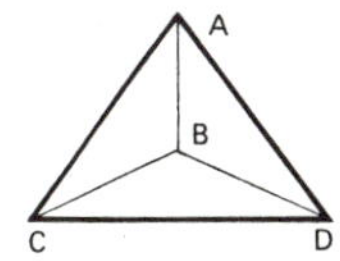


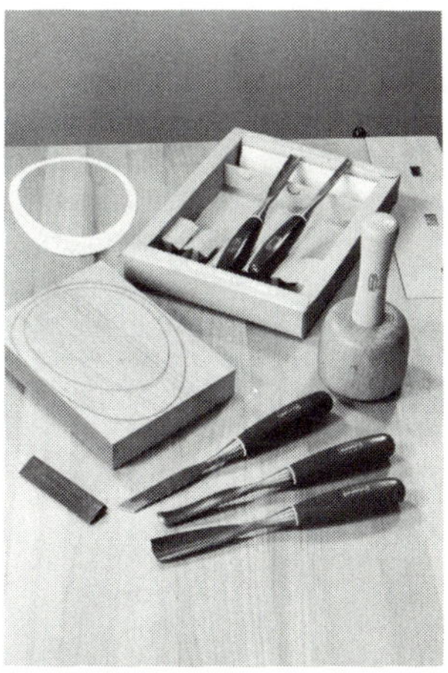

**TOOLS USED** are (from bottom): fantail gouge, bent gouge, skew chisel and wood mallet.

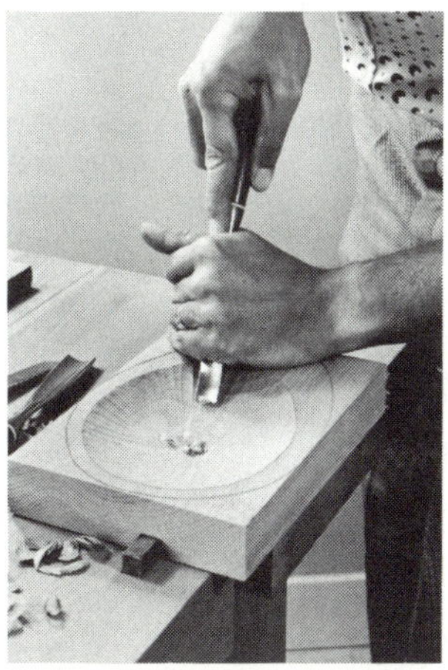

**THE RIGHT HAND** exerts pressure and the left hand guides as the bent gouge gives texture.

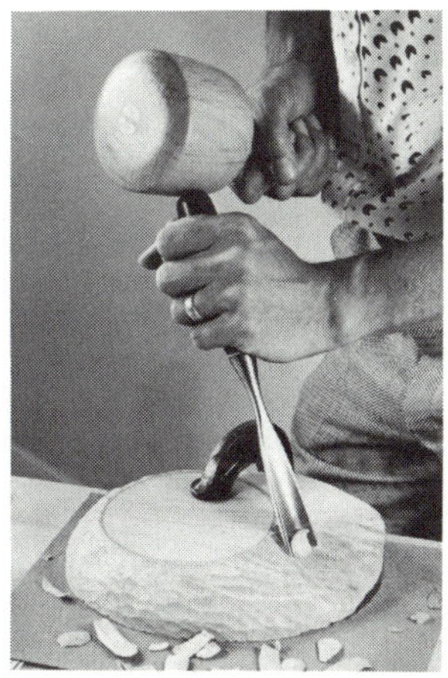

**THE WOODEN MALLET** exerts extra pressure on the fantail gouge as the dish is shaped.

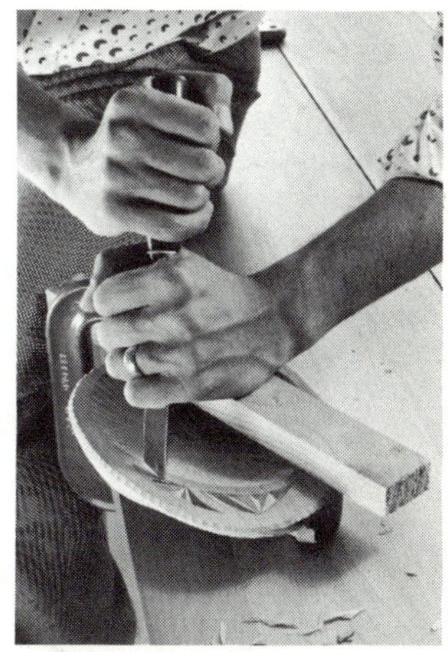

**THE SKEW CHISEL** is used for chip-carving of the finishing triangle decoration.

# 2 TRINKET BOX

**TOOLS USED** are (from bottom): straight gouge, bent gouge, skew chisel, mallet and slipstone.

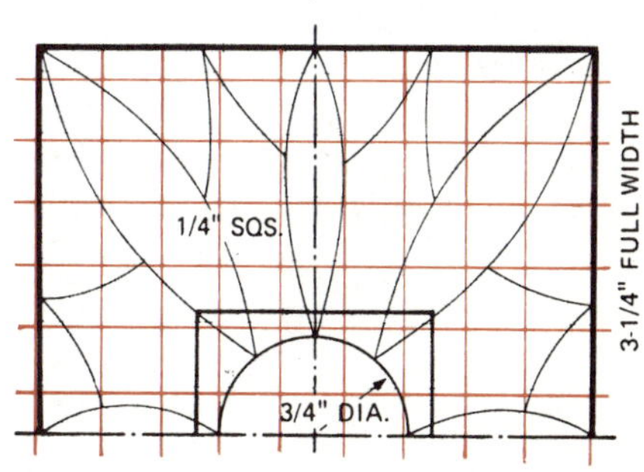


Here's a walnut trinket box with a low-relief flower design. For this you need a 4½ x 3½ x 8-in. block.

**Tools:** ⅝-in. flat straight gouge, ⅝-in. bent gouge, ⅝-in. skew chisel, mallet. 1-in. carpenter's chisel, bandsaw and coping saw.

**BOX LEGS** are marked and sawed to a ¼-in. depth. Here excess wood is being gouged out.

**INSIDE OF THE BOX** is hollowed with a bent gouge to a 1-⅜-in. depth, leaving a ¼-in. lip on the edge.

**SHAPE THE** three levels on the lid. Texture and beveled corners are made on the lowest level.

**THE FLOWER** design is drawn, then incised with a chisel; excess wood is then chiseled away.

**Method:** Make sure the wood is square, then with a circular saw or bandsaw halve it along the 4½-in. depth to make the box and lid. Go over the box with a flat gouge to give it texture. The box legs are ⅞ x 1¼ x ¼-in. deep. Mark off the first two dimensions and cut along marks to a ¼-in. depth with a coping saw. Remove excess wood with a carpenter's chisel or gouge. Mark inner space to be hollowed out, leaving a ¼-in. lip. With mallet and bent gouge, scoop out the interior to a 1⅜-in. depth.

Inside the box lid make a ¼-in.-wide by ⅛-in.-deep lip with a skew chisel. Find the center of the outside of the lid. Mark off a 1-in. square for the knob and a 2¼ x 3¼ in. rectangle for the flower petal area. Using a ⅝-in. straight gouge and mallet, level the area outside the rectangle sloping down to the corners. With the same gouge, level the petal area about ¾ in; draw the petal design.

With skew chisel, round the knob, incise petal lines and cut away petal background. Texture the lid with a flat gouge. Finish with oil.

# 3 PARTRIDGE MIRROR

**TOOLS NEEDED:** ⅜ and ⅝-in. flat gouges, quick gouge, chisel and mallet.

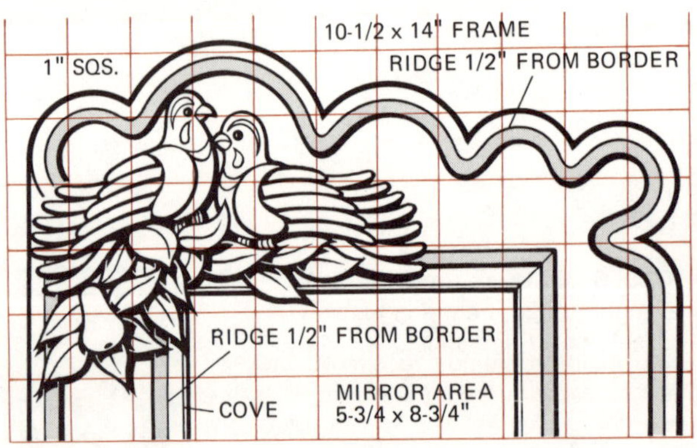


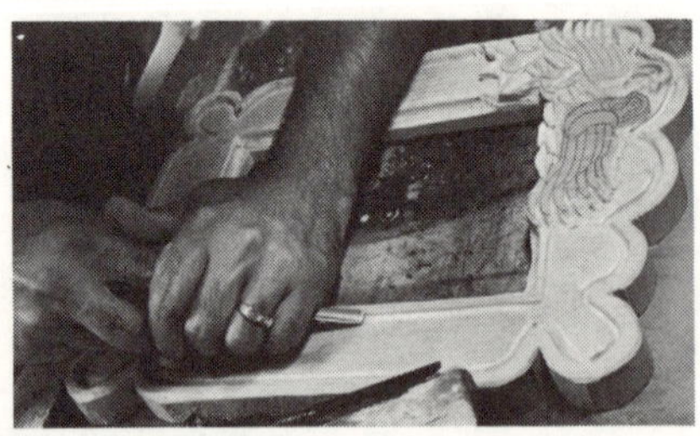

**MAKE** ridge ½-in. from inner border.

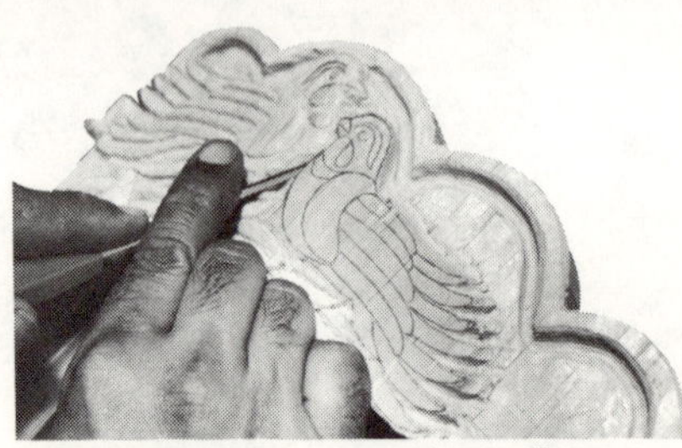

**CUT THE** design along drawn lines.

Our "partridges in a pear tree" mirror was carved from a single piece of white pine.

**Materials:** White pine board, 1¼ x 11 x 14½ in.; mirror, 6½ x 9½ in.; glazier's points or ½-in. brads; brown wrapping paper; white glue; hanging device; Minwax No. 224 Special Walnut Stain and Minwax Antique Oil Finish.

**Tools:** ⅜-in. flat, straight gouge; ⅝-in. flat, straight gouge; ¼-in. quick, straight gouge (remember "quick" means the end has a deep curve); ⅝-in. skew chisel; mallet; jigsaw.

**Method:** Begin with a smooth, flat board with square corners. Draw the frame including the 5¾ x 8¾-in. center mirror area on paper and transfer it to the wood. Saw the frame outline with a jigsaw. With the skew chisel, silhouette the birds. They are on the highest plane. Using the ⅝-in. flat gouge and mallet, reduce the depth of the rest of the frame about ¼ in. Remove the center mirror area with a jigsaw.

Mark off, then remove with a ⅜-in. flat gouge a ridge about ½ in. within the outer border and a second ridge ½ in. away from the inner border where the mirror will be placed. With the ⅝-in. gouge make a cove ⅛-in. away from the second ridge toward the mirror area.

Next, define the partridge detail with a skew chisel. Cut along the lines already drawn.

On the back, mark off ¾ in. around the inner frame perimeter for the rabbet that will hold the mirror and cardboard backing. With a skew chisel (you can cheat by using a router), remove the wood to accommodate the mirror and backing (½-in. depth).

After sanding the surfaces smooth, apply stain and oil finish to the frame.

Place the mirror and cardboard backing in the rabbet and secure them with glazier's points or ½-in. brads. Then dampen brown wrapping paper slightly larger than the mirror fame and glue it to the back. When the glue becomes tacky, trim the paper ¼ in. in from the border with a utility knife. As the paper dries, it will shrink and provide a smooth backing. Finally, add the hanging device.

**Hint:** You might first practice gouging and cutting scrap wood.

# 4 CAT

**TOOLS YOU NEED** for this project (from the bottom): 1-in. fantail gouge, ⅝-in. bent gouge, ⅝-in. flat gouge, ⅝-in. skew chisel, ¼-in. quick gouge and wood mallet.

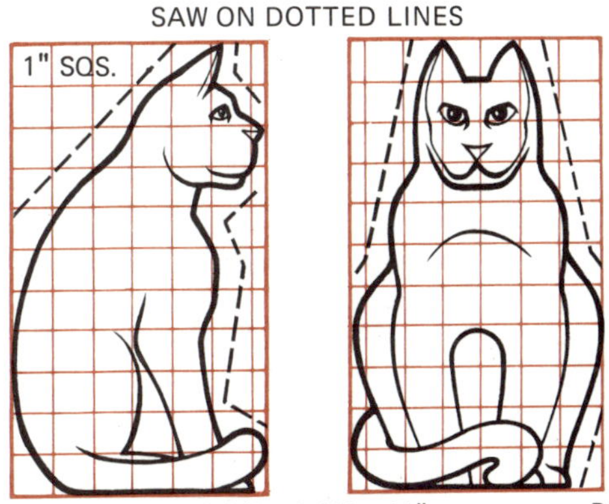


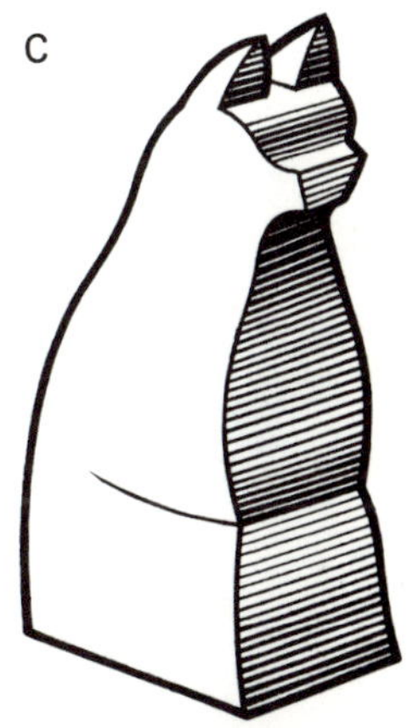


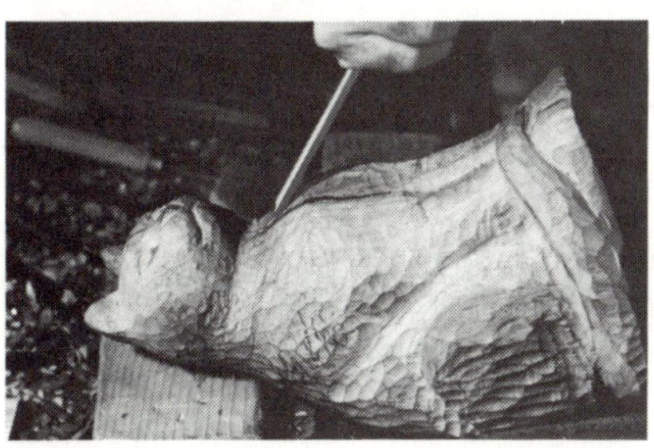

**WORK ON** the cat's neck is done with a ¼-in. quick gouge.

Carving in the round will test your ability to visualize a three-dimensional object. Our lifelike cat was carved from a 6¼ x 6¼ x 11-in. mahogany block.

**Tools:** You'll need a 1-in. fantail gouge; ⅝-in. bent gouge; ⅝-in. flat, straight gouge; ⅝-in. skew chisel; ¼-in. quick, straight gouge; mallet; wood rasp and bandsaw.

**Note:** You might get a clearer idea of how your finished piece should look if you make a clay model before you begin carving. It also helps to secure the bottom of your block of wood to the worktable with carver's screws, or by running screws through a piece of plywood and the wood block and clamping the plywood to the work surface. Doing so allows you to see all sides of the object and work on it in an upright position.

**Method:** Sketch the side views (A) of the cat on the wood block. Then cut away the waste with a bandsaw. Sketch the front view (B) of the cat and do the same.

As you work, it will help to draw in guide lines. With the 1-in. fantail gouge and mallet, continue to remove the excess wood and begin making rough details (C). A center line drawn down the cat's front will help you keep the proportions.

Begin to refine the details. The ⅝-in. bent gouge is good for rounding areas like the cat's haunches. The ⅝-in. straight gouge can be used to smooth the gouge marks made by the fantail gouge, and to further refine details. Use the skew chisel to outline and make

**A SMOOTH FINISH** is one key to producing this sleek, lifelike cat.

**WEIGHT** can be added to the mallet (above) by boring holes up to ⅜-in. diameter evenly spaced in the head and filling them with lead.

small details like the cat's mouth and paws. The ¼-in. quick gouge can be used to define small places like the neck, tail and between the legs.

Remember, whenever you must exert a great amount of force, your tools probably need sharpening—either with a slipstone or with both oilstone and slipstone.

Smooth the cat with a wood rasp followed by three grades of sandpaper. Apply three or more coats of linseed oil, allow 24 hours between coats, and finish with butcher's wax.

Often the simple action of a carving tool on wood acts as a burnisher. When the tools serve as a burnisher you may want to leave your work in its natural state without applying a finish. Before burnishing, strop the tools on a piece of leather containing a mixture of emery powder and petroleum jelly, immediately after they've been sharpened.

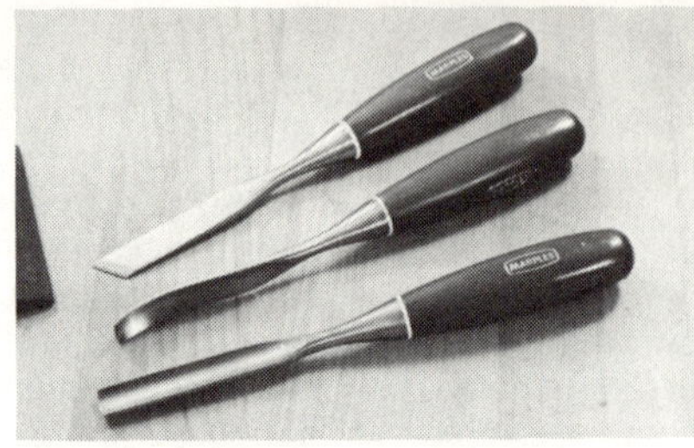

**AN ADVANCED** carver will need V tools and small gouges called "fluters."

Another simple way of burnishing a carving is by rubbing it with wood shavings.

### SHARP TOOLS ARE A MUST

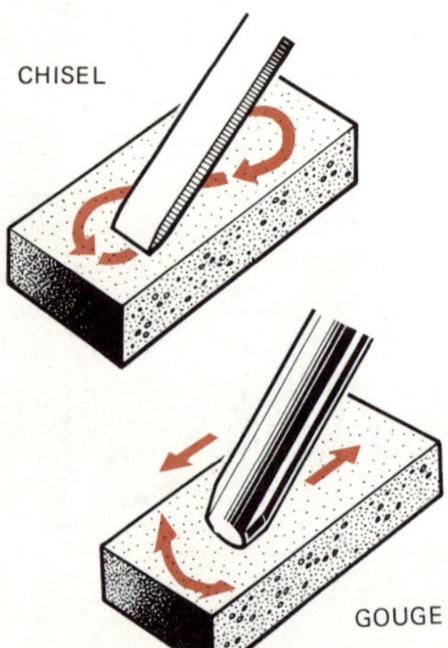


To sharpen new tools to the keenest edge and keep them finely honed, you'll need a medium/fine-grit sharpening stone with light oil and a slipstone. Sharpen the gouge by rubbing it back and forth on oiled sharpening stone, while you rotate it from edge to edge until a burr forms. Remove this burr with a slipstone, holding it flat on the inside of the gouge and square across the end. To sharpen both sides of the chisel, place each bevel on the oilstone and move the chisel in a figure-eight pattern. Continue honing each side until you break off a burr. After extensive use and a great deal of sharpening, these tools must be reground. Your local supplier most likely will have facilities for such work. If not, he could recommend a proper source.

### FINISHES FOR WOOD CARVINGS

Newly cut timber that will attract dirt and grease should be sealed. Several coats of white French polish applied with a soft brush and rubbed down with fine steel wool will seal it and leave a slight gloss.

Or you might choose to give a carving several coats of linseed oil. It can be burnished with beeswax applied with a coarse cloth.

The soft luster of wood doesn't lend itself to lacquering, varnishing or finishing with a method that gives a high gloss. But distressing of softwoods can make an interesting finish.

One method of distressing wood is to burn the surface slightly with a blow torch and later brush the wood in one direction with a steel brush. This treatment highlights the harder growth of the wood and improves the form of the object, but it should not be used with very resinous timbers or woods containing large knots.

Another method of distressing softwood is to poke holes in it with an awl or ice pick. This is generally followed by a dark stain for an effect like wormwood.

Use a good furniture polish occasionally to protect the finish.

### WOOD CHARACTERISTICS

- Basswood: A pale yellow, fairly soft, easily cut wood; not for detail work.
- Birch: A light-toned, fairly stable wood, but not too durable.
- Boxwood: Very hard; good for small, delicate work.
- Cherry: Fine working qualities; grain often adds to quality of carving.
- Chestnut: Similar to oak in appearance, but much softer. Tools must be very sharp to avoid crushing the fibers.
- Elm: Very strong grain which can enhance a carving. However, uneven grain can cause working difficulties.
- Lime: Excellent wood. Perhaps the best all-round wood for carving and sculpting.
- Mahogany: Various types available; American has good all-round qualities; cuts and polishes well.
- Pine: A soft wood, light in color and very easy to work. However, dull tools will crush and tear the fibers.
- Walnut: A dark, easy wood to work; it has even, close grain, cuts and polishes well.

**THESE POPPERS** produce good results. Here the author's son has used one to land a good-sized striped bass

**HEAD AND** tail ends are rounded on a belt sander. The shaft is smoothed with hand sanding

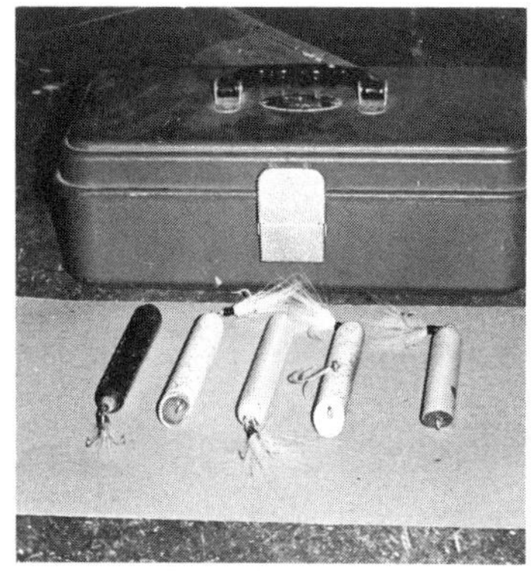

**BOTTOM HOOKS** were omitted on some plugs for easier casting and improved balance

**A HOLLOW** head for making the plug pop is easily routed into a concave cup with a spherical file bit in drill

**PLUGS ARE** painted several coats. One or two treble hooks are then placed in screweyes

**FINISHED PLUGS** are dressed with bucktail streamers of deer-tail hair tied on treble hooks

# Make your own poppers

By GEORGE LINNANE

WHEN MY SON cast the surface popper into his first school of bluefish, the strike was so vicious that it caught him unprepared. He fumbled with the antireverse lever and the blue lunged to the surface and broke off the line. It was hard to lose a king-size fish and even worse to see a $3 popping plug go with him. Homemade poppers might be the answer.

For basic material, we used ¾-inch birch dowels. An electric drill, mounted in an inexpensive drill press stand, and sanding discs are the only power-tool necessities, but since my shop has a table saw and belt sander, we used them as well. After studying a number of successful plugs, we decided one 4⅝ inches long with a ¾-inch diameter would match a seven or eight-foot spinning rod with 10 to 15-pound test monofilament line. Heavier lures for bigger tackle were later made of ⅞-inch stock in five-inch lengths.

First step is to cut a 5½-inch length of dowel. Next, drill one end with a ½-inch bit ½-inch deep. Then route this hole with a spherical rotary file bit, rounding it into the concave shape which causes the splashing and popping. We found this a critical operation that had to be done while the blank is held vertically with pliers in the drill stand. The blank can also be clamped in a vise and drilled freehand.

Before sanding, the hollowed end is cut at a 45° angle and then rerouted again with the rotary file bit to deepen the hole. The shaft can then be hand or disc-sanded. Weighting consists of drilling a ⅜-inch hole, ⅜-inch deep in the underside one inch from the tail end and filling it with molten lead. Total weight: about 1¼ ounces.

For painting, two coats of a flat oil-base primer proved necessary, plus three coats of acrylic spray or airplane dope. Metallic Christmas glitter was rolled into a final clear plastic coat. Three No. 214 ¾-inch zinc-plated steel screw eyes, one at each end and one 1¼ inches from the front lower lip are screwed into predrilled holes, with size 1 galvanized treble hooks secured to the back two. A white bucktail was tied to the tail hooks.

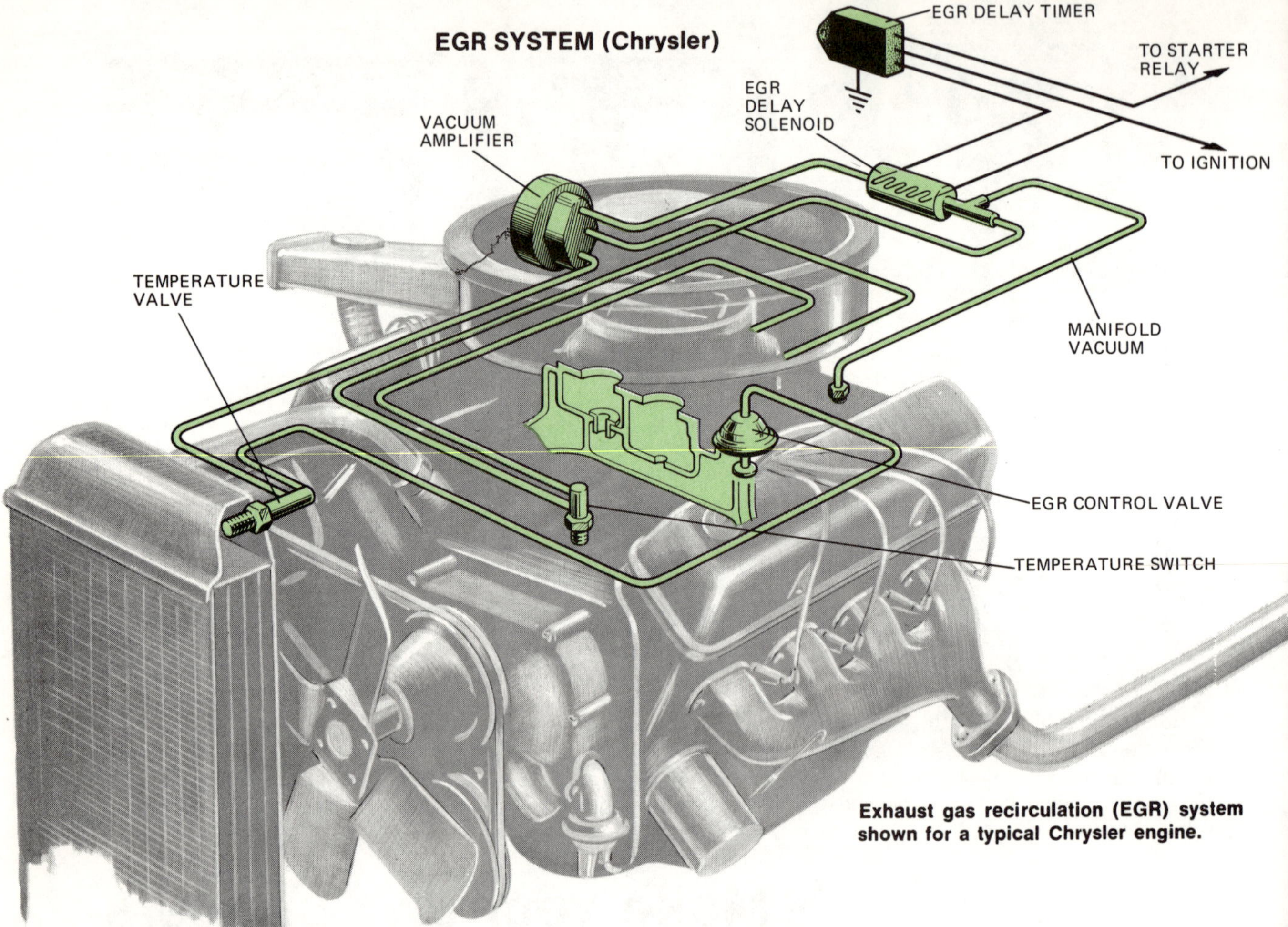


**Exhaust gas recirculation (EGR) system shown for a typical Chrysler engine.**

# How to service your catalytic converter and EGR system

By MORT SCHULTZ

A CATALYTIC converter, which is the muffler-shaped container hung under a car with the exhaust system to cut down on hydrocarbon and carbon monoxide emissions, has a life expectancy of 50,000 miles.

The EGR—shorthand for exhaust gas recirculation—is a means of lowering combustion temperature so oxides of nitrogen, pollutants that develop in a high-temperature environment, are kept in check. This is done by recirculating metered amounts of exhaust back into the combustion chamber, which slows the combustion process and lowers the combustion temperature.

The catalytic converter and EGR system are easy enough to keep in shape. And if you live in a state with mandatory emissions inspection, you don't have a choice.

SEE ALSO

**Autos, maintenance . . . Belts, auto . . .
Carburetors, auto . . . Chokes, auto . . .
Emissions controls . . . Engines, auto . . .
Exhaust systems, auto . . . Misfiring, auto . . .
Stalling, auto . . . Tune-up, auto**

### types of catalytic converters

There are two types of catalytic converters on U.S.-built cars. One is filled with thousands of alumina pellets. Alumina is an aluminum oxide. Pellets are coated with a combination of platinum-palladium, or platinum-palladium-rhodium catalyzing agent. The other, called a monolithic converter, contains a one-piece extruded honeycomb material that is coated with a platinum or platinum-palladium catalyzing agent.

Monolithic and pellet converters function the same way. The catalyst oxidizes exhaust from the engine into carbon monoxide (CO) and hydro-

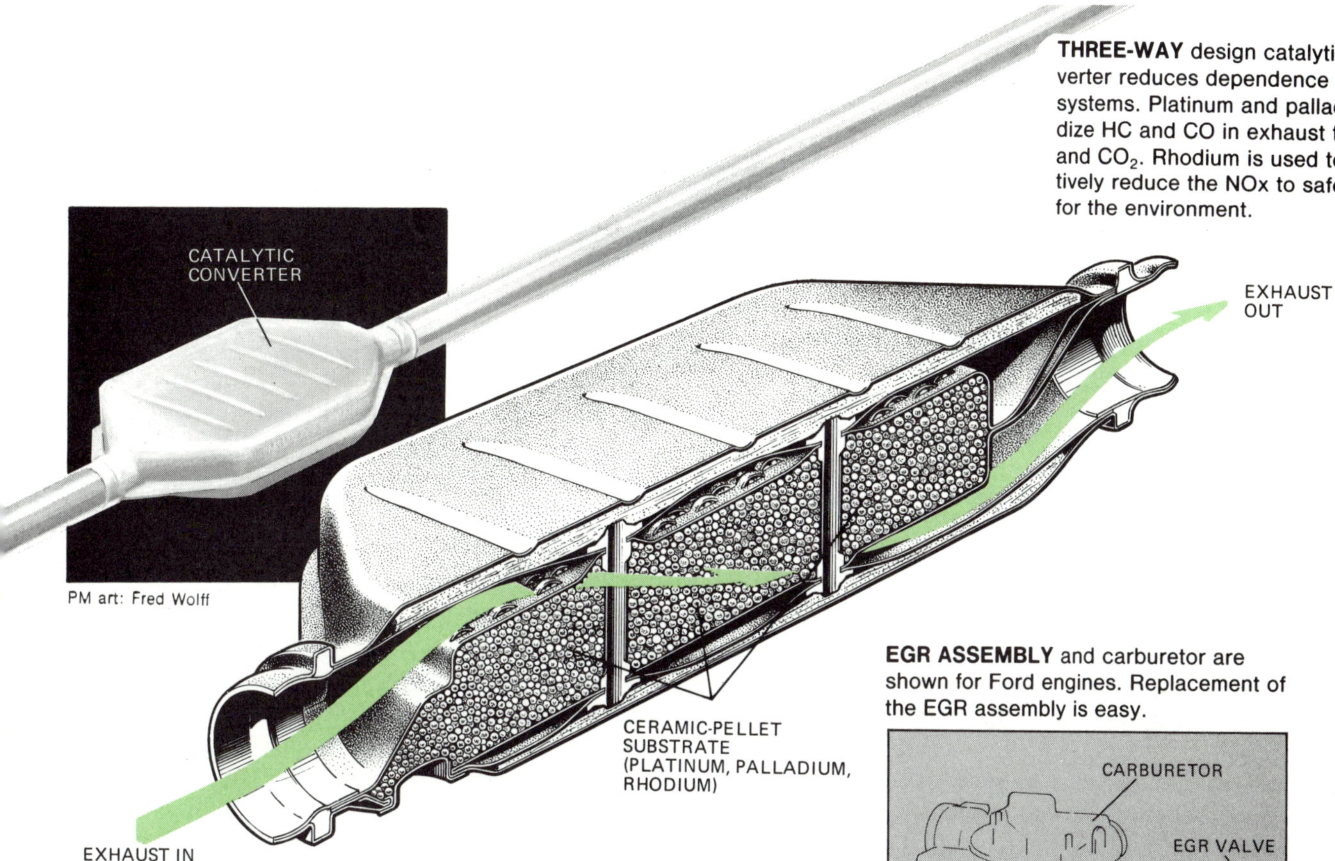


**THREE-WAY** design catalytic converter reduces dependence on EGR systems. Platinum and palladium oxidize HC and CO in exhaust to $H_2O$ and $CO_2$. Rhodium is used to effectively reduce the NOx to safe levels for the environment.

PM art: Fred Wolff

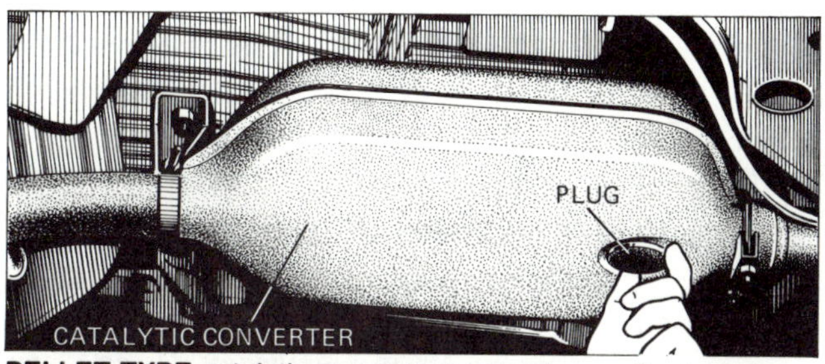


**PELLET-TYPE** catalytic converter has a plug on the bottom side. When the converter needs to be rejuvenated, only the pellets need to be replaced. Vacuum pump is used to suck out the spent pellets and then a vibrator tool jiggles the fresh pellets into the converter.

carbons (HC) into harmless water ($H_2O$) and carbon dioxide ($CO_2$).

In the process, the temperature inside the converter rises to a level somewhat higher than that of the exhaust gases. However, insulation inside the converter keeps the outside skin at about the same temperature as the muffler.

The main difference between pellet and monolithic converters, other than construction, is that the pellet converter can be serviced. Pellets can be replaced when they become ineffective. When the catalyst-coated honeycomb in a monolithic converter gives out the entire converter is replaced.

**EGR ASSEMBLY** and carburetor are shown for Ford engines. Replacement of the EGR assembly is easy.

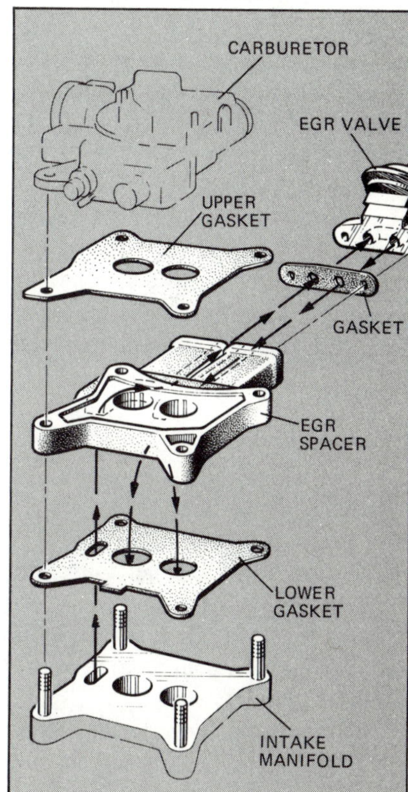


**EGR VALVE** diaphragm pulsations can be felt and are check of good operation.

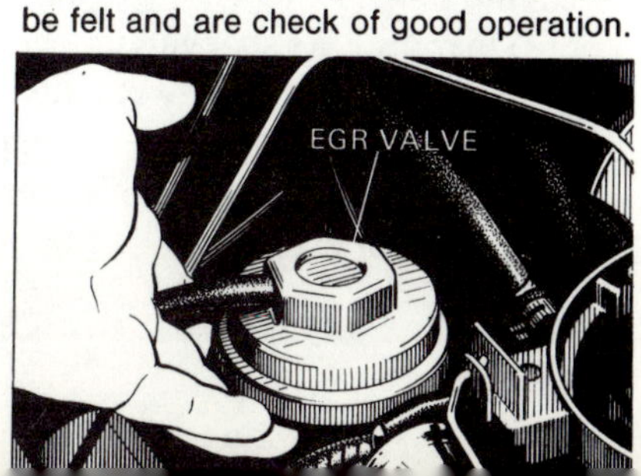

**replacement procedures**

Perhaps surprisingly so, replacing a monolithic converter is more of a do-it-yourself job than replacing pellets, because no special equipment is needed. The unit is replaced the same way a muffler is replaced. Ironically, however, replacing a monolithic converter yourself (or pellet converter, if necessary) is more expensive than having someone replace pellets.

If you have to replace an entire pellet converter the cost may be high, but this is necessary only when the outside body is distorted.

Distortion (bugle) occurs because excessive heat builds up in the converter—not the converter's fault.

Excessive heat is caused by a carburetor, ignition or air pump problem that permits unburned fuel to enter the converter where it ignites. If a damaged converter is replaced without correcting the problem, the new converter is going to fail prematurely.

If a pellet converter's bottom cover is ripped or dented, you shouldn't have to replace the whole converter, only the bottom cover.

Replacing pellets is done with a vacuum pump and vibrator—two tools that have the same part number (J-25077). They can be ordered from Kent-Moore Tool Div., 1501 South Jackson St., Jackson, MI 49203.

To replace pellets, a threaded or pressed-in plug is removed from the base of the converter. The vacuum pump is attached to suck pellets out and the vibrator, possessing a can of fresh pellets, is attached to pump the new pellets in.

**who's got what**

Although there are only two main types of catalytic converters, there are some interesting variations.

Some cars use pellet converters as the main instrument, but they also possess monolithic warm-up converters. Others have monolithic warm-up converters in combination with the main monolithic converters.

A warmup converter's small size and closeness to the engine cause it to react more rapidly than a main converter to incoming gases. It converts exhaust gases into nonpolluting agents when the engine starts.

Generally, a six-cylinder engine has one warm-up converter between the exhaust manifold and main converter. An eight-cylinder engine has two warm-up converters, one serving each bank of cylinders.

Replacing a used-up warm-up converter is as simple as replacing the main converter. Remove the front exhaust pipe at the exhaust manifold and at the pellet converter joint. The warm-up converter is integral with the front exhaust pipe.

A "phase two" catalyst emission control system controls oxides of nitrogen, as well as hydrocarbons and carbon monoxide (hence it is most-often referred to as a *three-way* catalytic converter).

It reduces dependence on EGR systems and ignition-spark retarding to meet oxides-of-nitrogen standards.

One of the catalyst coatings in a three-way system is rhodium, which controls oxides of nitrogen by separating this polluting agent into harmless nitrogen and oxygen. Unfortunately, rhodium is scarcer and more expensive than platinum and palladium.

The heart of the phase-two system is an electronic fuel control unit, which receives voltage signals from an oxygen sensor in the exhaust system. The sensor tells the control unit if the fuel mixture is too rich or lean. The unit makes the necessary correction at the carburetor. Accurate air/fuel ratio is critical to operation of the system.

The oxygen sensor should be replaced every 15,000 miles. Otherwise, servicing procedures for the phase-two converter are the same as for other pellet converters.

**beating an old horse**

By now everyone should know they shouldn't use leaded gasoline in a car that has a catalytic converter except in an emergency. Using a few gallons of leaded gasoline is acceptable to get a car to a gas station that has unleaded fuel.

Constant use of leaded gas will leave combustion deposits, mainly lead, in the converter, producing the same results as a blocked exhaust system—lack of power and limited top speed even at wide open throttle. According to Ford, rust that gets inside a converter from a rusting exhaust manifold will do the same thing.

**egr: the way it works**

EGR systems from car to car work pretty much the same. Technically, the system is called PEGR (proportional exhaust gas recirculation), because the design provides exhaust gas recirculation at a rate proportional to engine load.

No exhaust gas recirculation is provided at idle, because none is needed since combustion temperatures are low enough to restrict oxides of nitrogen formation. As acceleration is applied and the car reaches cruising speed, more EGR is provided as needed.

The job of metering exhaust gases to the air-fuel charge falls to the EGR control valve. The EGR valve, in turn, is controlled by vacuum.

In a typical engine, the vacuum signal is proportional to engine air flow (or load) and is provided by a port in the carburetor venturi. The vacuum signal is applied to a vacuum amplifier, which modulates vacuum in such a manner that output from the amplifier is normally 10 times the magnitude of the input vacuum signal. The amplified signal actuates the EGR valve.

There are other key parts to an EGR system besides the EGR valve—the coolant temperature switch and valve. These eliminate EGR when the engine is cold to provide stable engine operation during warm-up.

**testing the EGR**

If a part of the EGR system malfunctions, engine hesitation and/or poor idling result. The main components of a typical system—EGR valve, coolant temperature switch, connecting hose—may be tested as follows:

*Caution:* The engine has to be at normal operating temperature and the EGR valve may be hot. Be careful not to burn your fingers. You may wish to wear gloves.

1. With the engine idling, hold your fingers against the bottom (diaphragm) of the valve. You may feel a slight vibration—normal in six- and eight-cylinder engines.

2. Accelerate the engine. The diaphragm should open (move upward).

3. Return the engine to idle. The diaphragm should close (move downward).

*Note:* If the installation prohibits your being able to place your fingers on the diaphragm, watch the EGR valve stem instead. It should move as the engine is accelerated rapidly and is allowed to return to idle.

4. If the diaphragm (valve stem) doesn't move, pull the vacuum hose off the valve. Hold your finger over the end. Accelerate the engine to a speed equivalent to 2000 rpm. You should feel a strong pull. If not, remove the hose and check it for a restriction or cracks.

5. If vacuum is present, remove the EGR valve for cleaning. Or replace the valve.

*Note:* EGR valves are different from engine to engine of the same make. Be sure to get the right one.

6. If you are experiencing a driving problem when the engine is cold, and the EGR valve and hose test satisfactorily, test the coolant-temperature switch. With the engine cold, see if vacuum is present at the EGR valve by feeling for suction (Step 4). If it is present (it shouldn't be), replace the switch usually found at thermostat housing or coolant passage.

**specific service problems**

■ One reason for poor idling and hesitation for Ford engines can be a burned-out EGR spacer between the carburetor and intake manifold. This would allow a constant flow of exhaust gases into the intake manifold. Replace the spacer, and the upper and lower gaskets.

■ Some AMC four-cylinder engines use a stainless steel restrictor plate between the EGR valve and intake manifold. The plate and the two gaskets that sandwich it are calibrated for a particular engine-exhaust system combination and should not be altered or replaced with any other type of restrictor plate. There is an identification number or letter on every plate. Be sure the new part has the same number or letter as the old part.

**a troubling report**

One Environmental Protection Agency report emphasizes that emission controls are only as effective as we allow them to be. Analyzing the first substantial amount of emissions data from in-use vehicles showed that about 60 percent of 1975 models had emissions above 1975 Federal Standards at low mileage. The 1976 vehicles also showed a high failure rate. Reasons given were:

■ Misadjustment of basic engine settings, particularly idle speed, ignition timing and choke.

■ "Willful or inadvertent" disablement of a component or system. Examples include plugged, disconnected or rerouted vacuum lines; "carefully damaged" EGR valves; broken or missing idle mixture limiter caps.

■ Defective components, including leaking vacuum diaphragms, faulty coolant temperature-sensing vacuum switches and timers, broken EGR back-pressure transducers.

■ Inadequate maintenance (dirty air cleaners, worn-out plugs).

■ Using nonspecified parts. Replacement components should match the particular engine and emission control system combination. As an example, replace resistor-type sparkplugs with the same type, instead of using nonresistor plugs.

# How and what to caulk

■ THE HIGH PRICE of fuel is reason enough to seal your house tightly with caulk to keep out costly—and chilly—blasts of winter air. But there are other reasons for maintaining a good, periodic caulking program. A properly sealed home is protected from unnecessary damage—inside and out—due to the effects of weather extremes, wind, dirt and moisture. And caulking will do much to eliminate unsightly paint failures such as mildew and peeling.

A common mistake is to think that all caulk materials are alike. Believing that, many homeowners will buy a tube of caulking that sells for $1 or less—after spending $10 or more per gallon to repaint their homes.

A homeowner should be familiar with three types of caulk. Oil base is the cheapest. But, more often than not, the other types will do the job better. Here's the rule to follow when choosing caulk: Always pick the one best suited to the job at hand. The chart below will help you do just that.

SEE ALSO
**Bathtubs . . . Gutters and downspouts . . . Home improvement . . . Insulation . . . Windows . . . Winterizing, homes**

**What kind of caulk should you use?**

| Type | Advantages | Disadvantages |
|---|---|---|
| Oil base | Low cost. | Short use-life. Requires recaulking each year. Can't be painted for 24 hours. Flexibility is limited. Has minimal allowance for movement. Has poor adhesion to many surfaces. |
| Butyl | Long life. Good to excellent adhesion to most surfaces. Good flexibility at most temperatures. | Solvent required for cleanup. Composition tends to be stringy. Also tends to pick up dirt. |
| Acrylic latex | Long life. Good to excellent adhesion to most surfaces. Requires only 30 minutes setup time before painting. Easy to handle. Good weathering characteristics. Water and soap used for cleanup. | Should not be subjected to rain immediately after application. |

**STOP COSTLY DRAFTS** by caulking joints where siding meets door casings. Clean out old caulk, and make certain the surfaces are clean and free of all dust. Apply a bead, holding the caulk gun at a 45° angle to the surface. Your best bet: *acrylic-latex caulk.*

**SEAL OPEN JOINTS** between house siding and foundation. If crack is more than ½ in. deep, push in urethane filler before you begin caulking. Squeeze caulk from the tube and force it into crack with a putty knife. *Use butyl or acrylic-latex caulk.*

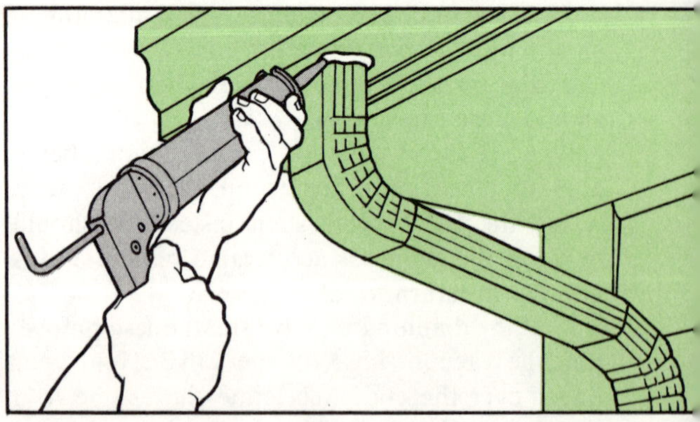

**LEAKS APPEARING** at rusted drop outlets where downspouts join gutters can be stopped by caulking. For good adhesion, be sure surface is completely dry and free of rust. After applying caulk, tool bead flat with a putty knife. *Use butyl caulk.*

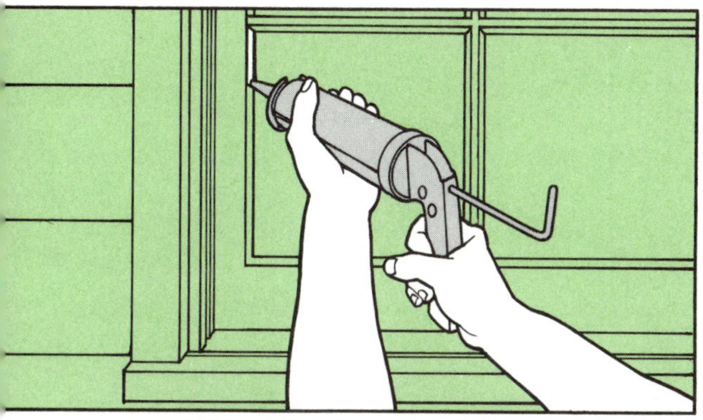

**CAULKING MATERIAL** makes quick work of reglazing chores. Remove all loose chips, dust off and apply caulk to the glass-wood joint. Spread the caulk with a putty knife; periodically dip the knife in water to make the spreading easier. *Use acrylic latex caulk.*

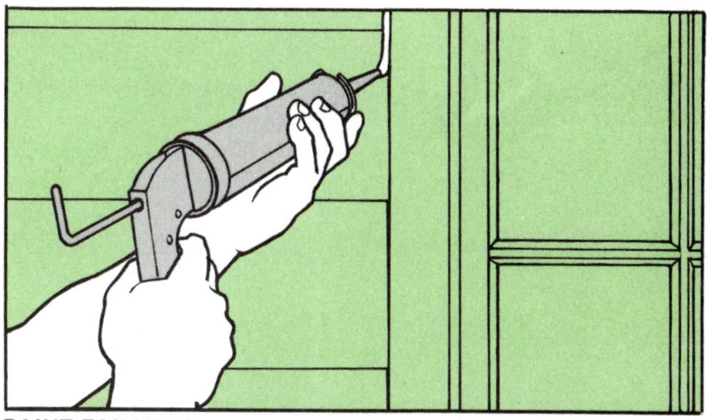

**PAINT FAILURES** are usually due to moisture entering siding material at the end grain; thus, joints at siding /window-casings should be caulked. Other spots to caulk: Where two lengths of siding butt each other and at the corners. *Use acrylic latex caulk.*

**CRACKS AROUND** wall caps (for ventilators) and airconditioners should be sealed with a bead of caulking between the siding and the fixture. If desired, cap can be loosened and then reset in bed of caulk. *Use acrylic latex or butyl caulk* for this job.

**UNSIGHTLY CRACKS** or nailholes in siding can be filled quickly by forcing caulk into the crevice, then feathering the edges smooth. After removing any excess caulk, the material can then be painted to match the rest of the siding. *Use acrylic-latex caulk.*

**TO PREVENT WATER** from seeping into your home, seal every plumbing fixture (hose bibb shown) and electrical outlet. Force the caulk into the joint around fixture. If necessary, first stuff in urethane or some other nonstaining filler. *Use butyl or acrylic latex caulk.*

**LOOSE FLASHING** around chimneys can be a source of continual water leakage into the house. To close this gap, run a bead of caulking into the seam between the flashing and shingles. If necessary, also caulk between flashing and chimney. *Use butyl caulk.*

**CEILINGS** now become an important part of the decorating scheme instead of being plain overheads that go unnoticed.

**THIS ARMSTRONG** tile, called Headliner Weatherby, is a good example of the new decorator tiles, featuring a textured motif on a textured base.

# How to install ceiling tile

**A tiled ceiling can be an easy do-it-yourself answer if you want to control noise in your house, lower a ceiling, or just put up a beautiful, room-warming topper**

**By LEN HILTS**

SEE ALSO

**Additions, house . . . Attic dormers . . . Bars . . . Basement remodeling . . . Basement stairwells . . . Bathrooms . . . Bearing walls, removing . . . Contractors . . . Entryways . . . Family rooms . . . Garage remodeling . . . Hardboard . . . Home improvement . . . House additions . . . Insulation . . . Measurements . . . Paneling, hardboard . . . Paneling, plywood . . . Paneling, wood . . . Plasterboard . . . Remodeling . . . Remodeling, exterior . . . Room dividers . . . Scaffolding . . . Siding . . . Sliding glass doors . . . Storage ideas . . . Storage walls . . . Tile, floor . . . Tile, wall**

DO YOU HAVE a ceiling problem? If so, there's a good chance that you can find a solution through the use of ceiling tile.

Because of the simple installation systems available today, you can do the work yourself and end up with results equal to anything you might pay to have done. The work is simple and straightforward, requiring no special tools or techniques. In most cases, you can do a room from start to finish over a weekend.

Which of the following problems do you have?

**Noisy room.** If you have a room—for example, a family room—with a noise level that is sending you out of this world, install an acoustical ceiling. The little bouncing decibels won't ricochet off acoustical tiles, and the noise level will be appreciably lower.

**Cracked, damaged ceiling.** If the thought of renovating a cracked, damaged ceiling by replas-

**THE PLANKED** ceiling is another new design for tiled ceilings. This one is random width and length oak planks in the color of natural wood.

**THE SQUARES** are accentuated in this ceiling with rough fabric-surfaced tiles. Big variety of designs gives unlimited decorating latitude.

tering or installing new drywall is scaring you, relax and check your dealer for a ceiling tile design you like. You can cover the old ceiling in a day.

**High ceilings.** Do you have one of those marvelous old houses with 10 or 12-ft. ceilings? If so, you are wasting money pumping heat into all those cubic feet of empty space above the standard 8-ft. height. The logical way to improve the room dimensions and save on energy is to suspend a new ceiling at the standard level.

**Dull, uninteresting room.** You've had painted walls and painted ceilings and a dull, unspectacular room for a long time, and now you'd like to give it some pizzazz. There are scores of ceiling-tile answers for this problem. If you have bypassed tile in the past because you didn't like the "tile look," go back to the store and look again. In addition to an amazing range of textured and sculptured tile surfaces, you'll now find designs in cork, wood paneling, marble and wood planking. With these available, you can take off in dozens of interior-design directions.

**Insulation problems.** Insulation is one problem you shouldn't try to solve with ceiling tiles, and it is mentioned here only to save you from spending money on an application that will be ineffective. First, you shouldn't install insulation between two heated rooms, such as a family room in the basement and the kitchen upstairs. You really want an exchange of warm air between those rooms. Second, the R value of ceiling tiles is too low to make much difference. And third, true insulating panels, such as those made of glass fibers, are more effective when installed in walls and roofs, in normal insulation installations, rather than in a ceiling.

**Room light problems.** If you have rooms that are dark or need a new lighting treatment, once again, look to a tiled ceiling. The majority of ceiling tile systems include a variety of matching translucent panels which can be intermingled with the tiles. Result: Built-in lighting, strategically placed to provide an even light over the entire area.

**FIRST STEP** in installing Armstrong Easy Up tile system is to nail metal tracks to the old ceiling. Special nails come with the kit.

**NEXT, POSITION** the tile against the track and insert its tongue in the groove of the previously installed tile. System works for planks and tiles.

**A new ceiling.** If you are adding a room, or finishing a basement or attic area, you need to install a ceiling. One way is to use drywall. Another is to put up a tile ceiling. The tile installation is quicker and easier, and allows for easier light-fixture installation.

**what's involved?**

To install a tile ceiling in your home you first must select and buy the tile, along with the installation system which goes with it. Here is some buying information:

**Tiles:** You can purchase tiles in standard 12x12-in. tiles, 24x24-in. panels, and in 24x48-in. panels. Some systems may even offer other sizes, such as 12x48-in. Which size should you buy? In part, this question will be answered by the design that appeals to you the most. The larger panels offer a wider variety of unusual designs. Another consideration might be that you can finish a room faster with larger panels because there are fewer pieces of suspension track and fewer panels to put up in any given room.

The biggest and perhaps most difficult decision will come in deciding which of the wide array of designs you want. Your major consideration in this decision should be the use and decor of the room. Some of the designs are formal and elegant—hardly the thing for a child's room or the laundry area. Others are simple and not intended to catch the eye. Would you like the majesty of a marble ceiling? In the right room, a marble ceiling would provide great eye appeal and a lot of conversation.

Or perhaps you might consider a ceiling of wood. You can find a variety of wood effects including square oak panels and oak or pine planking designs. If the room is already paneled or you intend to panel it, you can install a ceiling to match or to provide a warm contrast.

When inspecting tile designs, remember that not all ceiling tiles absorb sound. If sound control is one reason for installing a tile ceiling, then be sure the design you select is made to absorb sound. Many designs have virtually no acoustical effect. Those that are acoustically designed are clearly labeled, so check the label before buying.

If you are planning to install a tile ceiling in a kitchen or bathroom, check the labels and literature for information on cleaning. Some wood fiber tiles cannot be washed, only vacuumed. Others can be sponged with mild soap and water. Some acoustical tiles can be painted a number of times and still retain their sound absorbency,

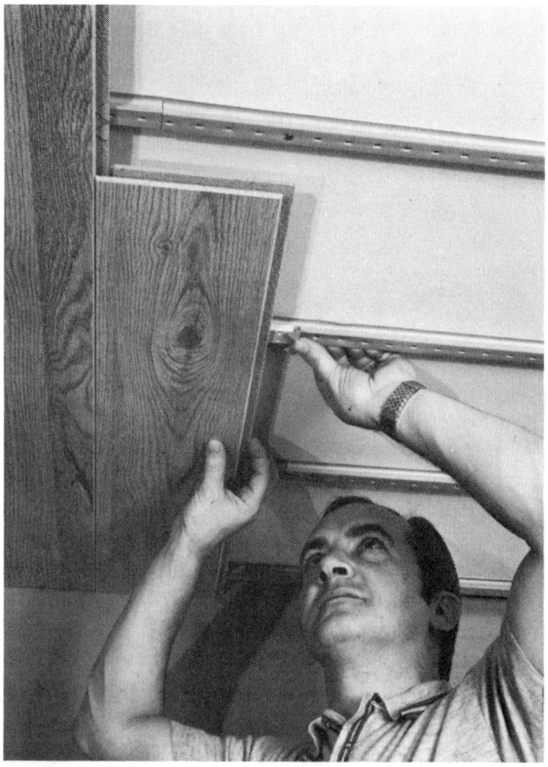

**SPECIAL CLIP** snaps on the track and slides over the tile lip. Clips are hidden when tiles are up, and only one clip per tile is needed.

others can be painted a few times, and still others cannot be painted at all.

If you are putting a ceiling in the kitchen, figure that you'll have to clean it often. Even a good exhaust fan won't prevent airborne grease from being deposited on the kitchen walls and ceiling, so cleaning is a real necessity. In bathrooms, steamy showers cause moisture to form on the ceiling, and the moisture deposits minerals which will have to be cleaned eventually.

Other rooms, such as bedrooms or living and dining rooms, get dirty at a much slower rate, and the dirt is easier to clean off. All of this means that any tile put up in "dirty" rooms such as the kitchen should be washable, while tile in the other rooms probably will look good with simple vacuuming and damp-cloth wiping.

One final word on design. Some tile systems are designed to emphasize the seams where the tiles join. In others, the metal parts of the installation system show as a part of the ceiling design. Others are made so that you see no seam or metal at all. Again, the choice is yours. But if you are allergic to the "tile look," the no-seam systems should appeal. In many of these, only you will know that the ceiling is tiled.

### tile systems

There are three basic systems for installing ceiling tile. In the first, the tiles are cemented directly to the present ceiling. In the second, wood or metal furring strips are attached to the ceiling, and the tile is either stapled to the furring strips or supported by them, depending on the system. In the third, the entire ceiling is suspended on a metal frame that hangs beneath the original ceiling suspended by wires.

There are a lot of variations on these systems, since each maker has developed his own version. In simple suspension systems, you can use any maker's tile as long as the tile panels are the right size and cut square at the edges. In simple furring-strip installations, you can install any flanged tile, since you staple through the flange into the furring strip. But when you look at any system employing clips or other special devices, be careful because these are a part of an integrated system that probably won't interface with any other system.

The important point to remember is: Be sure to buy a whole system—tiles and suspension parts—or carefully check to see that the tiles and other parts you buy will mix. Note that one manufacturer may make two or three systems, and

**ANOTHER METHOD** for installing tiles on a sound ceiling is to staple them in place. Use long staples and do not staple to damaged spots.

**CEILING SYSTEM** by U.S. Gypsum uses tracks nailed to or hung below joists. Most systems are similar, but fittings are seldom interchangeable.

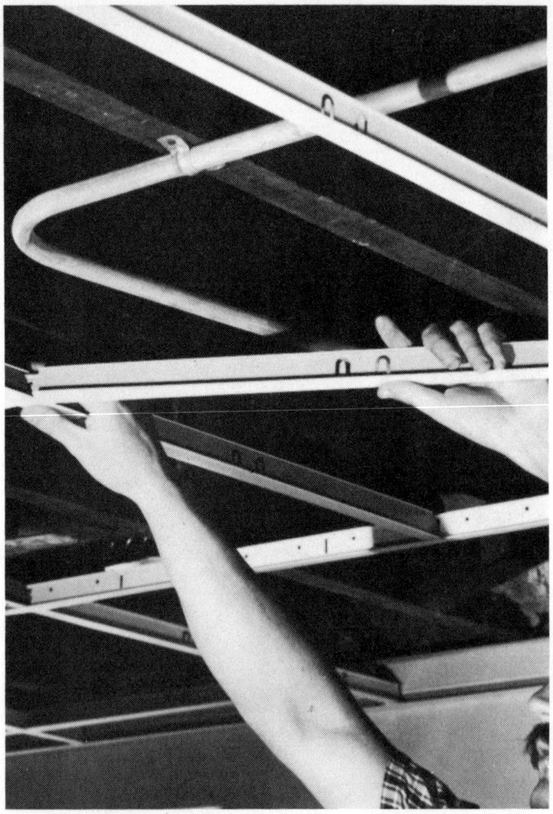

**ONCE THE TRACKS** are up, cross members are installed between them. Tile systems generally are light in weight and easy to install.

these systems may not mix—so don't buy by manufacturer's name alone. Look for the name of the system—such as Easy-Up, by Armstrong—and buy everything in that one system.

**System considerations.** The system you select in part will be determined by the problem you are trying to solve. If you are putting up a ceiling as a finishing touch in a basement playroom, then you'll find floor-to-ceiling height a prime consideration. Most basement ceilings are less than 8 feet from the floor to the bottom of the floor joists overhead, and you'll want your ceiling to take up as little space as possible. In this case, the furring-strip method is best. You can use kiln-dried furring strips, to which you staple the ceiling tiles, or nail up metal tracks which support the tiles.

If you are lowering a high ceiling, you will need a suspension system, consisting of metal tracks that support the tiles. The tracks usually are suspended by wires at a number of points across the room. The wires attach to screw eyes you screw into the old ceiling.

To cover a damaged ceiling, you also should use the furring-strip method. In this case, you nail either wood or metal furring strips through the old ceiling to the joists above.

If you are putting a new tile ceiling over a sound old ceiling, then the easiest method is to cement the tiles directly to the old ceiling. You put a small dab of cement on each corner of the tile and add another dab in the middle. Then you squish the tile against the old ceiling and slide it into contact with adjacent tiles you have already put up.

### tile installation

You'll get a complete set of instructions with the tile system you buy. Follow the step-by-step operations as directed, but here are some general points.

**Furring strips.** When buying 1x2 lumber for use as furring strips, ask for kiln-dried wood and select straight unwarped pieces. Warped or bent strips will result in an uneven ceiling. The instructions with your tiles will tell you to establish centerlines across the room in each direction, and put up furring strips beginning at the center and working towards the walls.

Locate the ceiling joists and nail through the furring strips into these. The furring strips are

**WITH THE SUPPORT** members in place, installing the ceiling is just a matter of placing the tiles on the flanges of the framework.

nailed in rows located far enough apart so each tile is supported on two sides. You may have to shim out some furring strips to level them.

Tiles are attached to furring strips with 9/16-in. staples driven by a staple gun. Don't staple tiles to drywall or plaster surfaces. They won't hold. If you use metal-channel furring strips, you also will mount a wall moulding around the room to support the tiles at the wall.

After the furring is up, begin putting up tiles in one corner of the room. With wood furring, staple the first tile, then fit the flanges of the next tile into it. Push the second tile into a close fit, be sure the edges of the tiles are squarely aligned, then staple the second tile. With metal furring, slide the first tile into position, then insert the cross tee which locks the tile into place. (The use of cross tees varies with the system.)

You seldom will have the number of tiles in a row come out even, and the last tile in the row must be cut to fit. Follow the instructions for tile cutting, since they may vary with the tile material. Two things are important. Measure carefully before cutting, and cut carefully to make neat, square edges.

For suspended ceilings, screw eyes are installed at intervals in the present ceiling, and a wall molding is nailed up at the height of the new ceiling. The main runners which support the ceiling tiles are suspended from these screw eyes by hanger wires. Once the runners are suspended in place, the tiles or panels are laid in on their tile-supporting flanges, beginning at one corner of the room and progressing from there. Some systems use cross tees and some do not.

The installation of any of these systems proceeds quickly, so there is seldom any problem in doing a room in a day or over a weekend.

One final note. Order your tile delivered several days before you intend to put it up. Open the boxes and keep them in the house. This permits them to adjust to the atmospheric conditions of the house. Most tiles expand and contract slightly as the moisture content of the air increases and decreases. This also is the reason why most tile manufacturers instruct you not to press tiles together too tightly as you mount them. If they are slightly loose, they have room to expand.

**ELEGANCE** is the word for this formal dining-room ceiling. This design is a far cry from the plain tiled ceilings of the past.

**WARMTH AND CHARM** of 200-year-old beams is captured with replicas of expanded urethane.

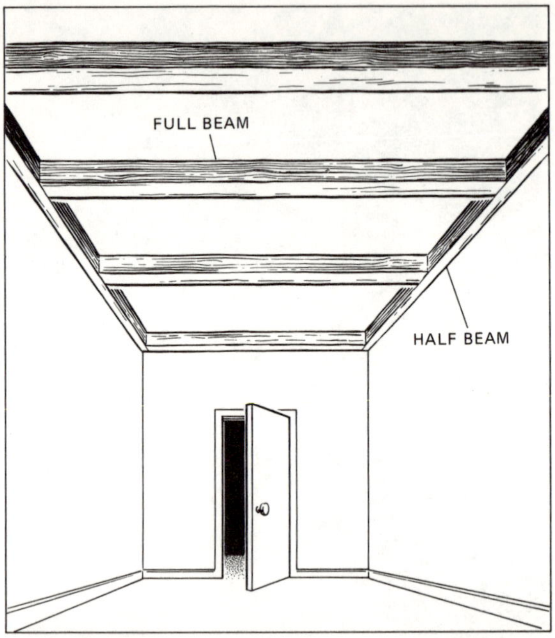


**BEFORE CUTTING** any material, you would be wise to make a scale drawing of the room, indicating beam locations. If the room is long and narrow, run the beams crosswise to provide the illusion of a wide room. The apron around the room, at the wall-ceiling angle, is created by using half-beams to make it seem that the other half is *in the wall*.

# Simulated ceiling beams— easy to install

SIMULATED "hand-hewn" wall and ceiling beams offer one of the easiest ways to add decorator elegance to an interior remodeling job. Available prefinished and ready-to-install, the easy-to-work beams can even be cut with a bread knife. They're applied using adhesive, nails or a clip-system designed by the manufacturer. Unfinished beams—which can be stained or painted as you prefer—are available by special order.

Ceiling beams come in 8 to 18-ft. lengths (in 2-ft. graduations); wall beams are sold in 8, 12 and 16-ft. lengths. The accessories available for use with the beams include matching corner braces with 45° mitered ends, recommended adhesive for best results, matching stains to disguise any marks, and simulated hand-hammered brackets for covering beam joints.

SEE ALSO

**Basement remodeling . . . Home improvement . . . House additions . . . Plasterboard . . . Remodeling . . . Sabre saws . . . Table saws**

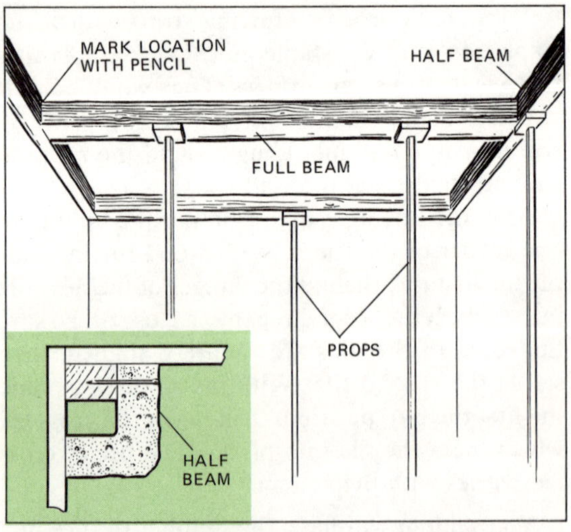


**SNUG BUTT JOINTS** are assured if you cut beams one at a time and temporarily prop them in position with thin sticks wedged between floor and beam (use block of wood under beam to prevent damage). To install half-beam apron permanently along ceiling line, first nail furring strip in the angle, then use adhesive and 8d finishing nails.

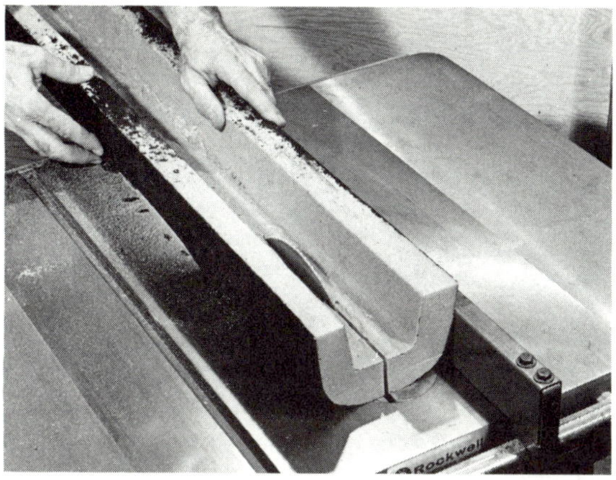

**A TABLE SAW** equipped with a fine-tooth combination or plywood blade is ideal for cutting urethane beams. If you don't have this equipment, the cutting will still be easy with a sharp, crosscut handsaw (and square)—it just takes slightly longer. To rip a U-shape beam, run it upside down (above) for safety and sufficient blade protection.

**BEAM CAN ALSO BE RIPPED** with a sabre saw after tacking a wood strip to plastic for use as fence.

**WHEN CROSSCUTTING,** if blade won't cut beam with one pass, make initial cut and turn over to complete.

**TECHNIQUES** for attaching plastic beams: When using self-sticking beams (above), you just peel off tab and stick into place. If installing beams to be positioned between joists (drawing below), strike centerlines where beams will be applied, and then install suitable-width furring strips using toggle bolts. Beams are then secured to the furring strips with adhesive and 8d finishing nails. When applying adhesive with a caulking gun (photo, below), add it sparingly down the center of the beam surfaces to avoid squeeze-out on ceiling or wall when you press beams in place. Finally, set finishing nails, fill heads.

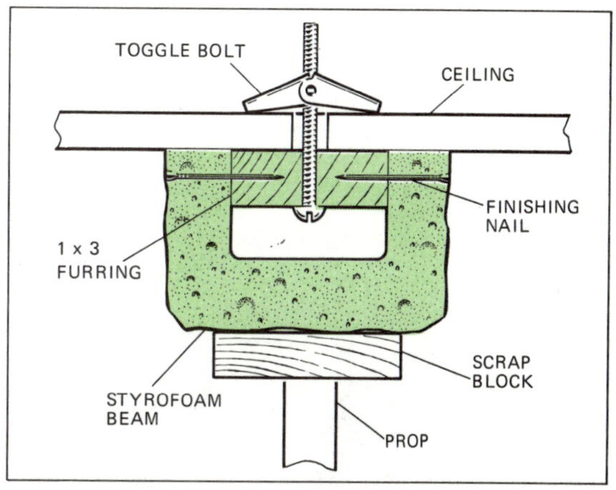

**BEFORE BEAMS** were installed, living-room ceiling looked as uninteresting as any other.

**AFTER BEAMS** went up, they served to direct attention to fireplace area, tied room together.

# Make your own antique beams

**By BERNARD POWELL**

**Using a homemade adz, you can add the ancient art of hand-hewing to your woodworking skills. Here's how one man did it, employing ingenuity and salvaged timbers**

SEE ALSO

**Adhesives . . . Basement remodeling . . . Calipers . . . Finishes, wood . . . Home improvement . . . Sanding . . . Scrapers . . . Staining, wood**

DURING A RECENT house redecorating session, my wife decided that what we needed to complement our colonial decor was a beamed ceiling. Though I was aware of the new foam-type beams that are good reproductions, I decided that I would carve out my own.

To make them, I used rough-sawn, nondimensioned stock measuring about 4x6 in. that I obtained at no cost. Actually, the beams shown were flatcar dunnage (pine) salvaged from a railroad siding. The railroad people were happy to

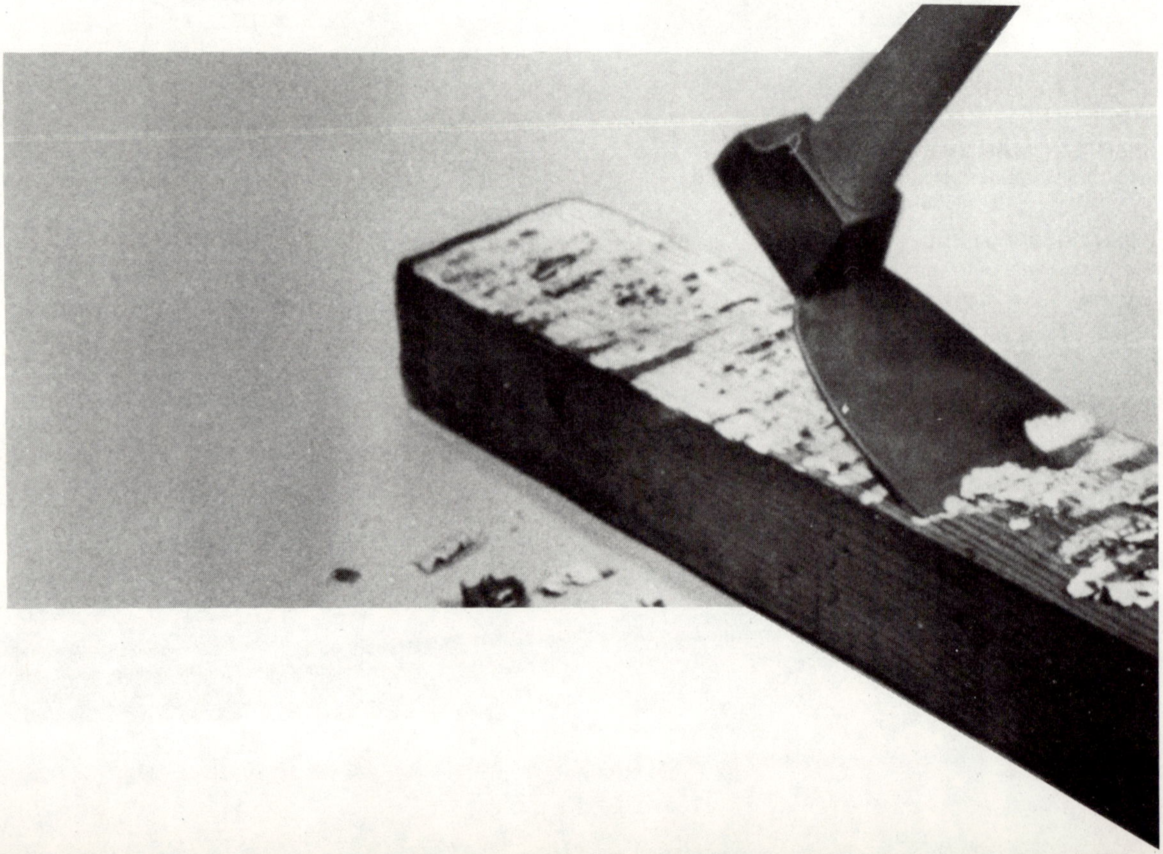

**SURPLUS PICK-MATTOCK** is rough-ground into an adz, using portable grinder. Bevel is on the inside.

**CUTTING EDGE** is honed razor-sharp to insure fine cuts and ripples that mark the work of an adz.

**PLUGS ARE CUT** from side that butts against ceiling, and are used to fill holes that came with timber.

**COATED WITH GLUE,** the plugs are then driven into enlarged nail holes before adzing begins.

**EASIEST METHOD** of adzing is to straddle raised beam. Use workshop or clamps to prevent movement.

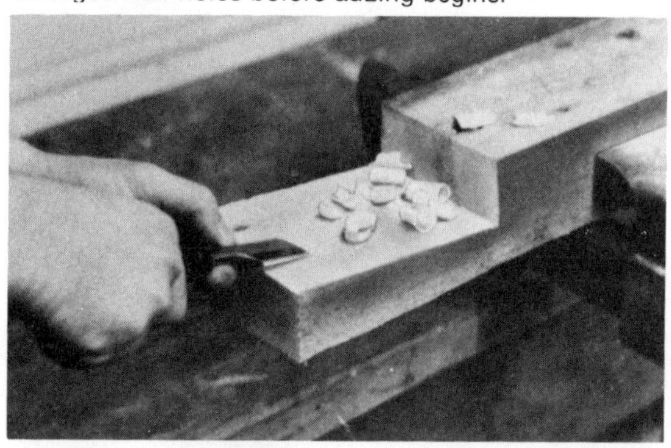

**TO MAKE UP TIMBERS** for long spans, beveled lap splices are cut on ends of beam to be joined.

have me haul the timbers away. If you have no local railroad freight yard, you can substitute dimensional lumber.

**make your own adz**

Though the adz is a tool that has been in use for centuries—it was the only known method of dressing lumber—it is not easy to come by these days. (And the chances are, if you do locate one, the price will be high.) Thus, my first problem was, how could I achieve an adzed look on my beams? I tried an ax, hatchet and a spoon chisel. None gave satisfactory results or the look that I wanted. The obvious answer was to make my own adz.

The dictionary describes an adz as a cutting tool having a thin arching blade set at right angles to the handle. (The tool in the lead photo is a genuine adz.) So, I picked up a used pick-mattock (the mattock blade fits the description) at an Army surplus store for a couple of bucks. Then, I ground a razor-sharp edge on the tool. After

**BEVELED SPLICES** are checked to insure fit, then assembled, using glue, lagscrews and dowel-plugs.

**DATE CARVED** in main beam is optional but adds to realism. Or, use the year you bought the house.

doing over 200 feet of beam hewing with this homemade tool, I can testify that it works fine.

### how to use an adz

To hew, you straddle a timber that is held securely by workstops on the floor. Use short, well-aimed strokes down and towards you in the curvature of the blade. Your arms and the handle act as the radius. The blade is worked across the width of the beam stroke by stroke, removing several inches of stock in length and width with each chop.

This is tricky. An angled stroke, or a hard chop that strikes a knot, can glance off and give a painful and serious cut in the shin. To avoid this, many pros working at a fast pace stand in wooden nail kegs to protect their legs. Your safest bet is to work slowly, always making certain that the cutting edge strikes *with* the grain on each stroke. When working over a knot, use much shorter strokes or a chisel.

Having a shorter handle than an adz, the mattock-adz shown on page 662 is best handled by elevating the workpiece and following the safety rules mentioned above. Do one side from end to end, then rotate the timber 90° and do the next side. Always work with the grain; cutting against it leaves deep, ugly gashes. Your first efforts may result in cutting sharp, jagged splinters. These can be removed with a hand chisel.

### installing the beams

Since the room where I planned to beam the ceiling had a wide entry, I ran a wide beam

**BEAM END** is chiseled out to receive wall-knee support. Kerfs made with a power saw speed this step.

**KNEE IS ATTACHED** permanently to main beam with glue and dowels before the beam is installed.

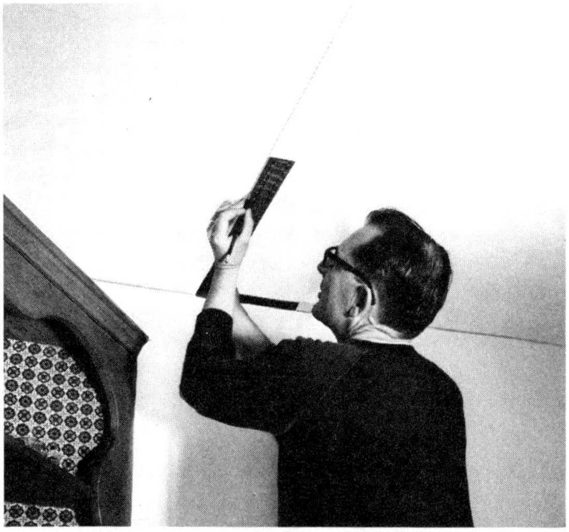

**CEILING JOIST CENTER LINE** is established on both sides of room and chalkline is snapped across.

across the opening. The long-spanning beams (actually joists in old construction) are at right angles to the heavy beam. They simply butt the walls at one end and are let into notches cut in the heavy beam at the other. The notches are cut extra deep so that the beam can slide in as you swing the far end up, thus letting the beam clear the wall.

Installation goes a lot faster if you have some help hefting the beams into position. And, it's worth the time it takes to build a T-support, which, with several wedges and a helper, makes installation comparatively easy.

Predrill and counterbore all beams for lag-screws and bolts while still in the workshop. Then lay out beam locations on the ceiling with ruler and chalkline. When a beam is positioned and wedged firmly in place, drill pilot holes into the joists to receive the lagscrews. After the beam is securely fastened, the shimmed T-support can be removed safely.

For extra support, I bolted knees to wall studs beneath each end of the heavy beam. About 18 in. long, the ones shown were found—as is—alongside the dunnage beams in the railroad yard. They had been discarded after serving as chocks under a large cast-iron boiler brought in on a flat car.

The beams shown were finished with walnut stain and two coats of semigloss varnish. This part of the job goes a lot faster if you do the staining and first varnish coat before the beams are in place. Then, it's a simple matter to touch up the few spots that might get bruised during installation and apply the second coat of varnish.

**PREDRILLED BEAM** is positioned over chalkline, held fast by T-support wedged with shims, then bolted.

**POST SUPPORTING BEAM** stays put until all bolts are turned home. Knee is fastened to wall next.

**WALL KNEE,** attached to beam, is positioned against the wall and drilled for lagscrews that will hold it.

# Musical centerpieces for Christmas

**This Christmas you can hear your own family singing carols, or Santa's voice under the tree or from the centerpiece. The trick? A cassette recorder**

IF YOU DON'T HAVE a cassette tape recorder, perhaps you should. You can make this Christmas a truly musical one by using it to provide personalized "built-in" sound in your table centerpieces.

The three examples shown are designed so that any cassette recorder may be slipped in place from the side or bottom. Cassette-type recorders can be set to play up to 60 minutes and, since you can make your own tapes, you can be as original and personal as you like. For example, the sound coming from the tree base can be the voices of your own family singing Christmas carols. Music emerging from the table candelabra can be a recording of Johnny or Susan playing *Silent Night, Holy Night* on the family organ. It can be Dad making out like jolly ol' St. Nick himself to keep tiny Jim believing in Santa. The tunnel-like candelabra consists of 13 layers which stand like slices of bread. Except for the center layer, there are two each of identical

SEE ALSO
**Candle stands . . . Cassette recorders . . .**
**Christmas cards . . . Christmas decorations . . .**
**Dining tables . . . Gifts, Christmas . . .**
**Lazy Susans . . . Toys**

overall size and shape. Both ends are alike also, except that the removable one has a centering plug glued to its back.

Since two pairs of each layer are alike, the U-shaped sides are bandsawed four at a time. This is done by first stacking the ¾ x 1½-in. pieces (all cut to proper length) in a block and holding them together with masking tape. As there is only one "A" layer, only two pieces would be stacked for cutting.

The U-shaped pieces are joined together at top and bottom with identical-length cross pieces. All are ⅝-in. wide so that when they are glued to the ¾-in. side pieces, ⅛-in. slots remain between the individual layers.

When all layers are completed they're glued together and clamped into a single block, taking care to keep all even at the sides. "Stringing" them on a scrap board will help to keep them aligned.

**The Styrofoam tree** has a base made up of five plywood discs, the bottom one being removable for placing a recorder inside. Sound comes out through a perforated metal band wrapped

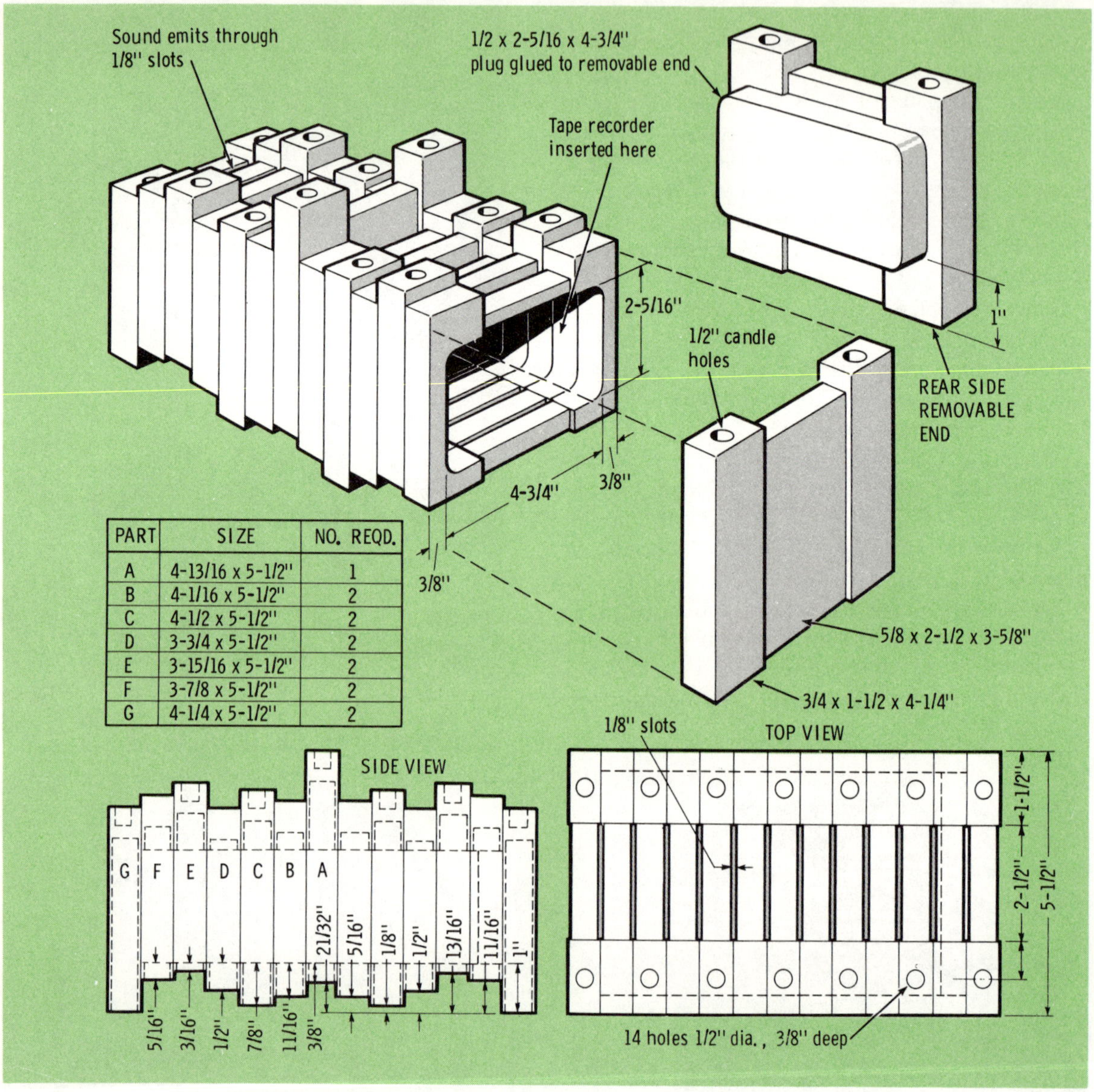


| PART | SIZE | NO. REQD. |
|---|---|---|
| A | 4-13/16 x 5-1/2" | 1 |
| B | 4-1/16 x 5-1/2" | 2 |
| C | 4-1/2 x 5-1/2" | 2 |
| D | 3-3/4 x 5-1/2" | 2 |
| E | 3-15/16 x 5-1/2" | 2 |
| F | 3-7/8 x 5-1/2" | 2 |
| G | 4-1/4 x 5-1/2" | 2 |

around two discs that are made slightly smaller in diameter than the top and bottom ones to which they are glued. The different size creates rabbets for the metal band. The base is completed by gluing and nailing ¼-in. thick wooden fins completely around the circumference.

The tree itself is formed in a triangular hollow pyramid from 101 Styrofoam balls tack-glued together with an electric glue gun. Building them up is tricky but can be done with help by assembling each side flat on a table and then tacking the three sides together on the inside with glue. The completed tree should measure nine balls on a side, nine balls high.

**The modern tree stand** shown on page 628 has an access door on one side. The details on page 631 show how it's made of two six-sided discs cut from ¾-in. wood and then joined with five beveled side members to form a hexagonal drum. Prior to assembly ⅛-in. saw kerfs, radiating to the six corners, are made in the top and down the edges. The side members are actually lined up with the corner slots so you have a continuous ⅛-in. slot from top to bottom. The sixth side member is the door. This piece and the two adjacent to it have holes drilled in it to let the sound out.

The assembly is supported by six L-shape legs cut from tempered hardboard. These, as well as the narrow strips for the top, are inserted in the slots during final assembly. Note that the legs also fit in slots cut in a slotted central base block, which is glued to the bottom.

The base is covered with six pieces of bright

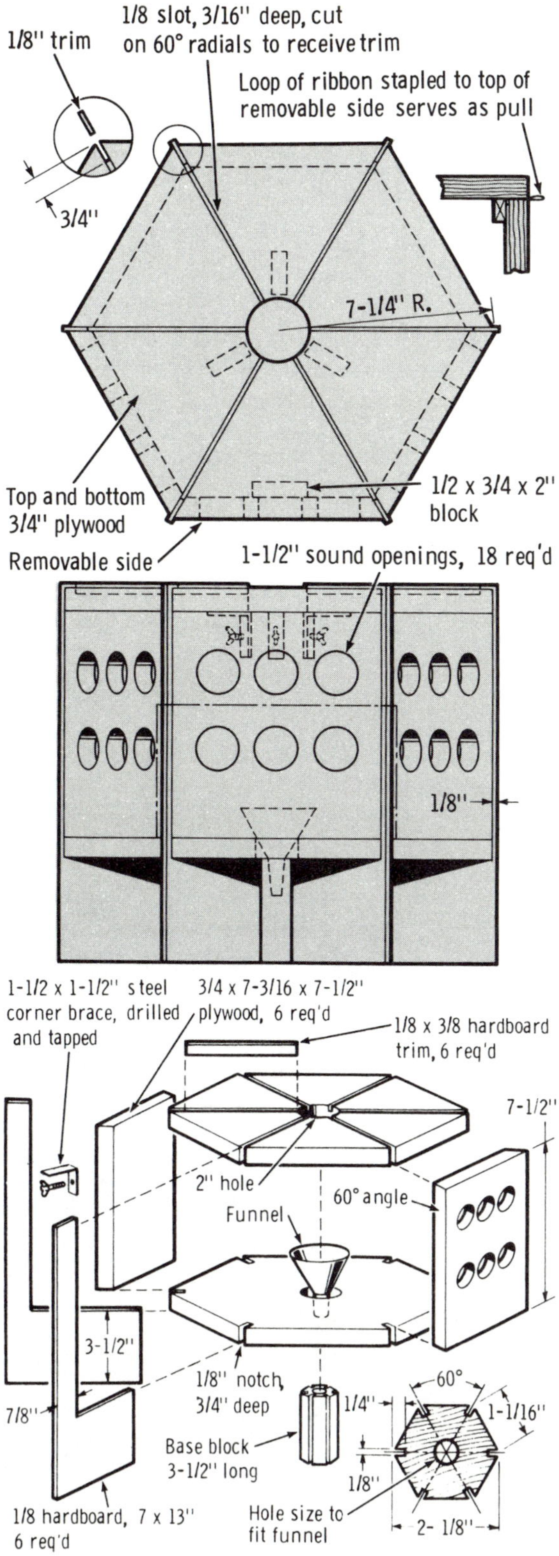


red felt, starting from the 2-in. hole in the top and bringing it down the side. These are cut to lap the slots slightly and are stapled at the hole and to the underside of the base. The hardboard legs and the narrow top strips hold the felt in the slots. In the case of the door opening the felt is tacked to the underside of the top. The door is covered separately with enough felt being left at the bottom to provide for tacking it to the underside of the base. This provides a hinge.

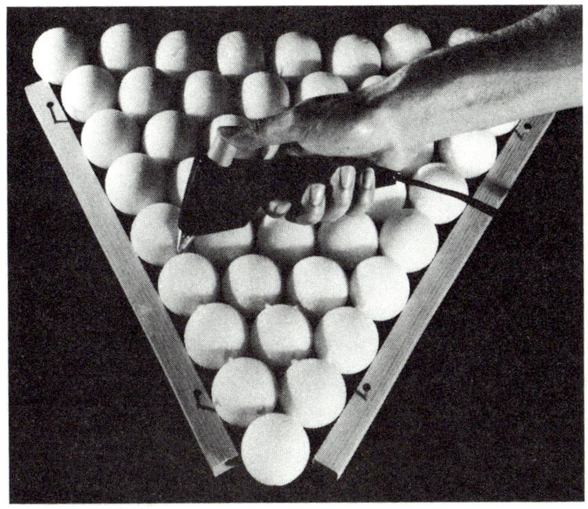

**HANDIEST WAY** to assemble the Styrofoam balls is to tack-glue them with an electric glue gun.

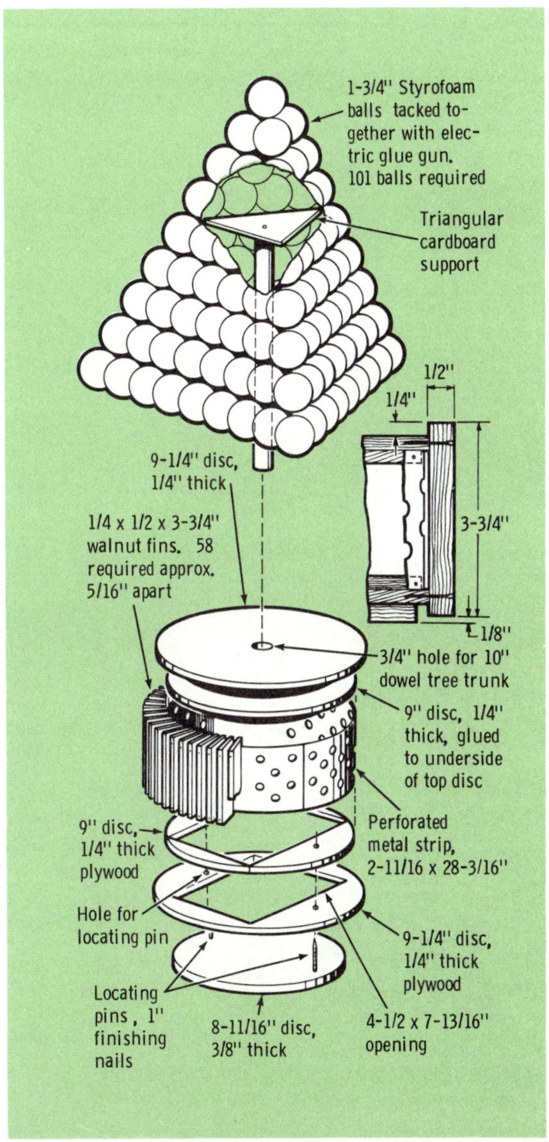

# Keep your chain

HAVING THE RIGHT saw chain on your chain saw and making sure it is kept sharp is the surest way of getting the most out of the tool. You can tell when your chain is getting dull by the size of the chips that fall at your feet. When the chain is sharp, the chips are large. A dull chain, on the other hand, spews out small, almost powdery sawdust.

A sharp chain will slide through a log with little pressure, whereas a dull chain requires excessive pressure, causing rapid wear of chain, bar and sprocket, as well as overloading of the engine. It takes only a few file strokes to keep a chain cutting efficiently, and saves you a lot of effort and aggravation later. The correct way to file a chain, plus other information on maintaining your saw, is covered in a booklet, *All About Using Chain Saws,* excerpted here. For a copy of the booklet contact Omark Industries, 9701 S.E. McLoughlin Blvd., Portland, Ore. 97722. And from now on—keep your chain sharp!

## HOW TO SELECT THE RIGHT SAW CHAIN

### MICRO CHISEL

| No. | Pitch | Gauge |
|---|---|---|
| 20 | .325″ | .050″ |
| 21 | .325″ | .058″ |
| 22 | ⅜″ | .050″ |
| 23 | ⅜″ | .058″ |
| 24 | ⅜″ | .063″ |
| 25 | .250″ | .050″ |
| 26 | .404″ | .058″ |
| 27 | .404″ | .063″ |
| 28 | .404″ | .050″ |

Fast, versatile with semisquare cutting edges. Easy to maintain with round file. A variety of sizes to fit most saws.

### CHIPPER

| No. | Pitch | Gauge |
|---|---|---|
| 102C | .250″ | .050″ |
| 45C | .354″ | .058″ |
| 12C | ⅜″ | .050″ |
| 51AC | .404″ | .058″ |
| 52AC | .404″ | .063″ |
| 50C | 7/16″ | .050″ |
| 61AC | 7/16″ | .058″ |
| 62AC | ½″ | .063″ |
| 9AC | ½″ | .058″ |
| 10AC | ¾″ | .063″ |
| 11AC | ¾″ | .122″ |
| 11BC | ¾″ | .122″ |

World favorite for over 20 years. Sizes for all saws, all timber. Fast cutting, easy sharpening. A real workhorse.

### CHISEL

| No. | Pitch | Gauge |
|---|---|---|
| 9AL | ½″ | .058″ |
| 10AL | ½″ | .063″ |

Square cutting edge never cuts same cross grain twice, removes more wood in a single pass. Requires skillful filing.

### OREGON S-70

| No. | Pitch | Gauge |
|---|---|---|
| 72D | ⅜″ | .050″ |
| 73D | ⅜″ | .058″ |
| 75D | ⅜″ | .063″ |

A good chain for lightweight compact saws. Cuts fast and smooth. Easily sharpened with round file.

### SUPER CHISEL

| No. | Pitch | Gauge |
|---|---|---|
| 50L | .404″ | .050″ |
| 51L | .404″ | .058″ |
| 52L | .404″ | .063″ |
| 50AL | .404″ | .050″ |
| 51AL | .404″ | .058″ |
| 52AL | .404″ | .063″ |

High-cutting efficiency and smoothness in both small or big timber. Easily sharpened with round file or chisel file.

### MICRO BIT

| No. | Pitch | Gauge |
|---|---|---|
| 58AC | .404″ | .058″ |
| 59AC | .404″ | .063″ |
| 68C | 7/16″ | .058″ |
| 69C | 7/16″ | .063″ |
| 31C | ½″ | .058″ |
| 32C | ½″ | .063″ |

A modification of the famous chipper design. It feeds considerably faster due to its cutting-edge profile.

### OREGON 80

| No. | Pitch | Gauge |
|---|---|---|
| 84 | .404″ | .063″ |
| 85 | .404″ | .050″ |
| 86 | .404″ | .058″ |
| 87 | .325″ | .050″ |
| 88 | .325″ | .058″ |

Fastest, smoothest chain made. Sharpens in seconds with Power Sharp system. Can also be hand-sharpened.

### GUARD LINK

Guards links are available on .325″ and ⅜″-pitch micro chisel and ⅜″ pitch S-70 (Speed guard).

Guard links provide increased protection from kickback when limbing and cutting in brushy areas. Good for use around home.

# saw cutting

## CHAIN-SAW TROUBLESHOOTING

Most chain-saw problems are caused by incorrect filing and improper chain tensions. Here are some common problems, their causes and remedies:

### Cutters won't stay sharp

| CAUSES | REMEDIES |
|---|---|
| 1. Incorrect filing angle (blunt or sloped. | Hold file correctly. Use a file holder. |
| 2. Depth gauges too high. | Lower the depth gauges uniformly. Use depth-gauge jointer of the correct size. |
| 3. Cutter lengths not uniform. | File all cutters to the same length. |
| 4. Chain not tensioned properly. | Adjust tension. |
| 5. Abrasive wear on cutters due to dirt and sand. | File away all abrasive damage. File undamaged cutters so that they match in length. |

### Chain cuts roughly, chatters or grabs

| CAUSES | REMEDIES |
|---|---|
| 1. Incorrect filing angles: feathered edge, hook, back-sloped, top angle not 35°. | File to specifications. Use a file holder and watch carefully to see that you are holding the file in correct position. |
| 2. Depth gauges set unevenly. | Use jointed and file uniformly. |
| 3. Improper chain tension. | Adjust tension correctly. *Caution:* When adjusting chain tension, be sure saw is turned off. If electric, see also that it is unplugged. |
| 4. Worn sprocket. | Replace sprocket. |
| 5. Rivet bearings worn out due to lack of oil. | Replace chain. Always use plenty of oil when cutting. |

### Chain doesn't cut straight (slanted cut)

| CAUSES | REMEDIES |
|---|---|
| 1. Saw not supported properly. | Hold saw with both hands and guide it through the cut. |
| 2. Cutters not filed uniformly; one side of chain uneven. | File all cutters to same length and angles. |
| 3. Depth gauges not uniform; some being higher than others. | Lower all gauges to the same setting; use jointer of the correct size. |
| 4. Abrasive damage to cutters on one side of chain. | File away damaged portions, then file undamaged cutters to match. |
| 5. Bar rails worn and spread. | Have bar rebuilt or replaced. |

## How to break in your saw

**Before starting** your saw, do the following:

**1.** Be sure the chain is mounted correctly, facing the right direction and properly tensioned. If possible, soak the chain in oil overnight before using it.

**2.** Fill the chain-oil tank with chain oil and check the chain oiler button to be sure it's working.

**3.** If it's a gas-operated chain saw, mix fuel and oil for engine in a clean container in the proper proportion. Use chain-saw oil recommended by the manufacturer. With an electric model, you can start cutting as soon as the tool is plugged in.

**To adjust tension,** loosen the bar stud nuts slightly. On hard-nose bars, tighten the tension screw until you can just pull the chain around the bar by hand. Hold tip of bar up while adjusting the tension screw and tightening the stud nuts. Otherwise, the chain will loosen on the first cut. Pump the oiler a few times, while pulling the chain around the bar by hand. Then start the saw and run the chain at fast idle for two minutes—without cutting! Stop the saw and recheck the tension. To check tension, let the chain cool and pump the oiler while pulling the chain around the bar by hand. Snap the chain in the bar groove to loosen and straighten all parts. If chain is loose, readjust tension.

**Start cutting**—only small limbs or logs for the first half hour. Use plenty of chain oil during break-in period and watch chain tension. A new chain will "stretch" during break-in and become loose. A chain can be ruined in five minutes of cutting if it is too loose. If it is too tight or slow cutting, excess heat and friction cause rapid wear.

## Type and size of saw

Most saws are of the direct-drive type today. There are gear-drive saws available, used mostly by professionals. The direct-drive saw is lighter, has fewer parts, requires less maintenance. It's easier to use, less tiring as the chain speed pulls the saw through the log without pressure. It cuts easily on any part of the bar.

For the occasional user, a lightweight saw is the best choice. Power is difficult to determine as there are no published horsepower ratings. The cubic-inch displacement of the engine is a rough guide to power—the larger the displacement, the higher the power. But there are other factors which affect the power output, such as compression, type of valving, and timing. The type of cutting chain, sprocket and guide bar are very important in the efficient use of available power. A small-pitch, fast-cutting chain running in a low-friction roller nose bar can effectively increase power. ★★★

# How to make tough cuts with your chain saw

By CHARLES E. RHINE

GETTING yourself and your chain saw up a tree, and felling limbs safely, takes know-how. This treeman is trimming small branches before cutting a heavy limb.

■ THERE ARE A NUMBER of reasons why a homeowner may have to remove a tree in crowded quarters. Perhaps it, or some of its branches, are so close to the house that it presents a hazard during storms. Or maybe some heavy, dead limbs pose a threat and should be removed. And, nowadays, trees that cast shadows where we plan to put solar collectors must go.

Before you can do anything, you must get up in the tree with a saw. Surprisingly, you can do this and still have your hands free for sawing by rigging up a rope saddle and a treeman's knot. See the drawing on the facing page.

This rigging allows you to safely control your up and down movements in the tree. If you slip, or a limb breaks under your weight, you won't come tumbling down, but just dangle until you can regain your footing. To free your hands for sawing, put your body weight in the saddle and brace yourself with your feet against the tree trunk or limbs.

The big secret to taking down limbs, or the entire tree, is to do it *a little bit at a time.* Saw off sections small enough to be easily handled and let down by rope. It takes time, but it's safe.

You need at least one other person on the ground to handle the ropes, and you need at least three strong ropes in good condition. Buy new ropes if there is any doubt about the condition of the ropes you own.

All ropes should be more than twice as long as the height at which you will be working in the tree. Rope lengths of 100 feet will handle most tree work. They should be free of knots and kinks and the frayed ends should be taped, not braided or knotted. Knots will catch in a tree crotch and prevent you from pulling the rope through.

One good way to test a rope for strength is to tie one end to your car's rear bumper and the other around a tree trunk. Ease the car forward until the rope is taut, then apply a steady pulling pressure. Look for areas of excess stretching. This is also a handy way to take the kinks out of a rope or to take the stretch out a new rope.

The most important rope is the one which will be supporting you, and with which you will tie your saddle. This should be at least ⅝-inch-diameter manila hemp rope—¾-inch rope would be even better. Nylon or other synthetic fiber ropes are too slippery for good control with the treeman's knot.

The second rope is the one you tie to the limb to be sawed off for lowering. It should be strong enough for the weight, so use ½-inch rope.

The third rope is used to haul your saw up after you have placed yourself in working position, or to swing branches out of the way of obstructions. A strong ¼-inch rope is fine.

Working up in a tree with a roaring chain saw can be extremely dangerous unless you have absolute control over the saw, your own movements and the movement of the limb you want to drop. If, at any time, you have doubt about your ability to do all three safely with a chain saw, by all means use a hand saw. At any rate, for best control, use only a lightweight chain saw.

If you try to use a chain saw, be sure to heed these safety hints:

■ Wear a hard hat, and if using a loud saw, ear protection.

SEE ALSO
Boxes, chain-saw . . . Crosscut saws . . . Engines, small . . . Fireplaces . . . Firewood . . . Sawbucks . . . Saws, buck . . . Trees

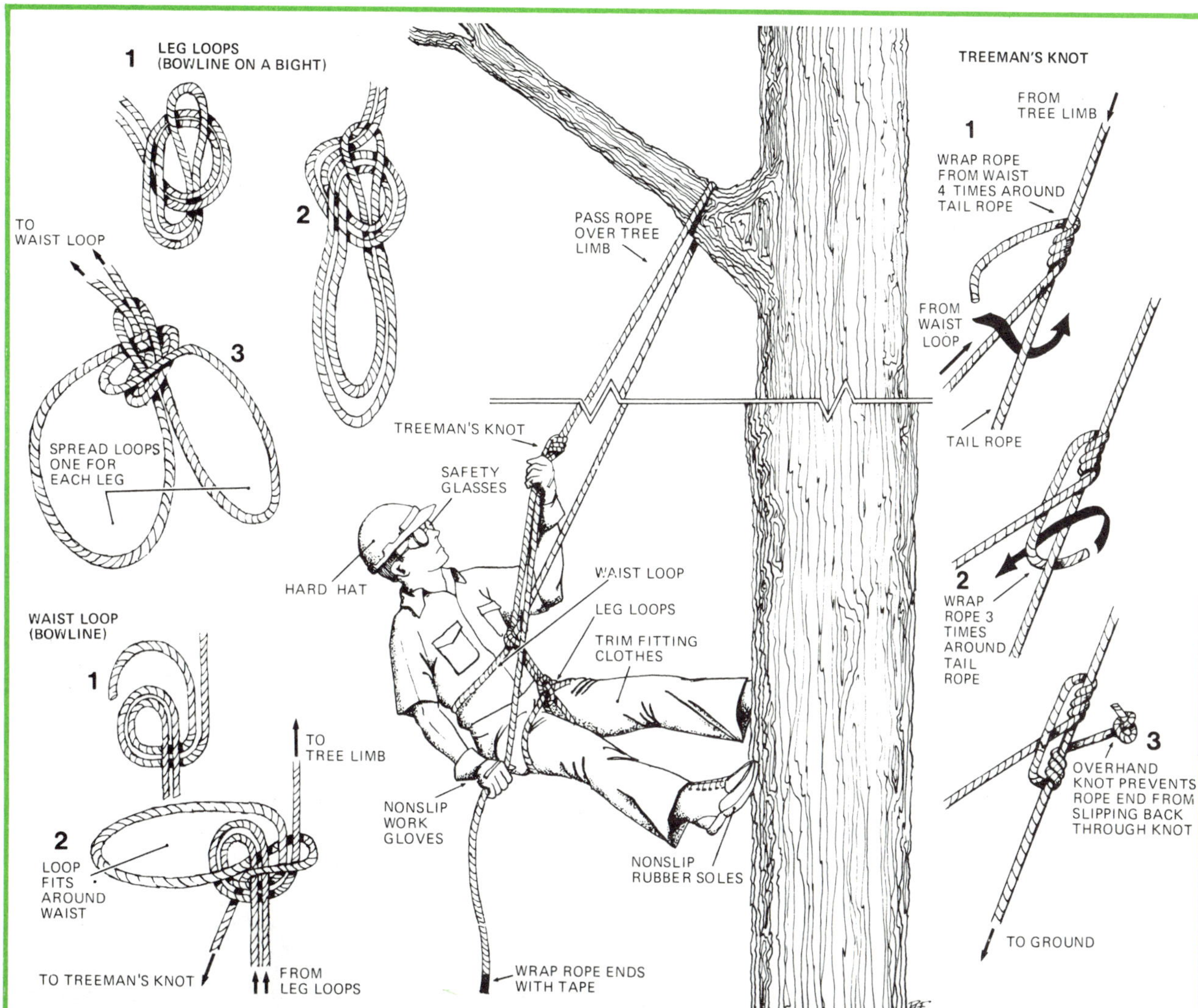


## WHAT YOU SHOULD KNOW ABOUT TREE CLIMBER'S RIGGING

**THE KNOTS** shown here make a safe saddle. You can become familiar with them through on-the-ground practice. They must be tied so they do not slip. Otherwise, they will cause the rope to tighten around you.

The saddle should have a snug, comfortable fit when your entire weight is on it. If the ropes bite into your skin too much, wrap them with rags or leather for padding. When you get to the site, it's a good idea to retie the knots a couple of times to get a good fit.

The saddle must be tied before you can pass the rope over the support limb. To start, measure off about 12 feet from one end of the rope and fold it back on itself. Begin tying your leg loops about three feet from the bend on the doubled end, using a bowline on a bight. Then proceed to the waist loop and the treeman's knot.

After passing the rope over the support limb, be certain that you test out your saddle and treeman's knot before you begin your climb.

**PASS ROPE** over strong, high limb after tying saddle and treeman's knots.

**BE SURE** to test your safety rigging near the ground before you climb tree.

**TREEMAN'S KNOT** suspends you safely, yet allows you to go up or down.

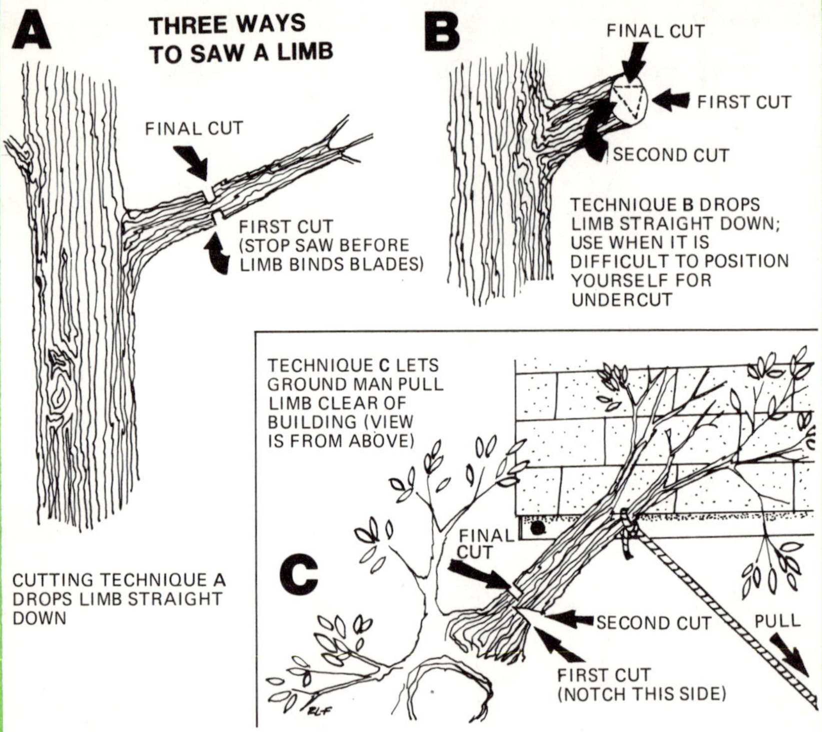


**CUTTING** technique (A) shows standard approach to dropping limb straight down from tree. Method (B) lets you make the final cut from above. All cuts should be in the same plane. Technique (C) enables the man on the ground to swing the limb clear of an obstruction or house.

### HOW TO THROW ROPE OVER A LIMB

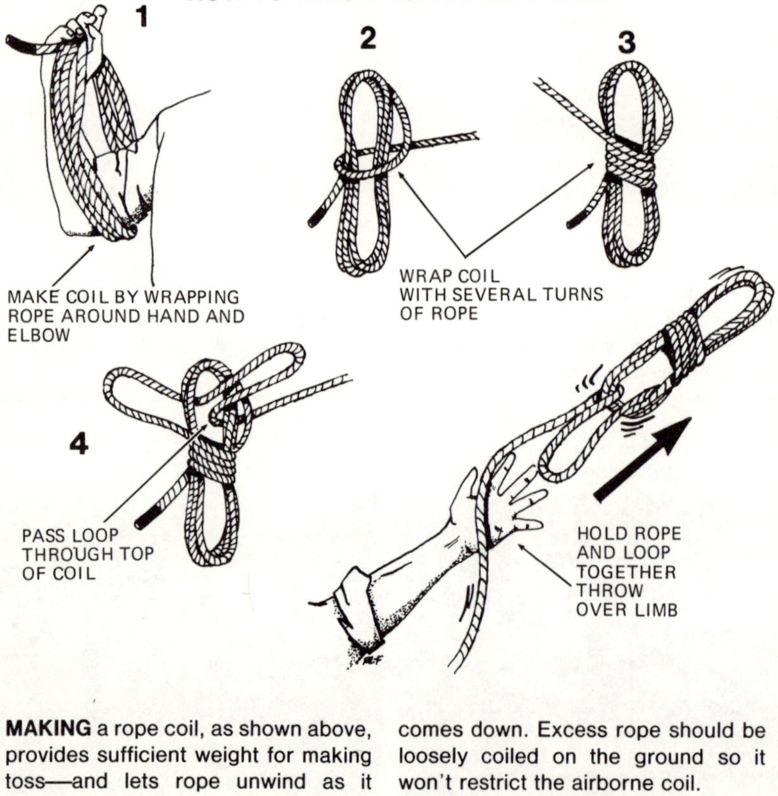


**MAKING** a rope coil, as shown above, provides sufficient weight for making toss—and lets rope unwind as it comes down. Excess rope should be loosely coiled on the ground so it won't restrict the airborne coil.

### HOW TO USE YOUR CHAIN SAW IN A TREE

**1 GROUND MAN** lifts saw to working position, using rope tied to saw's balance point.

**2 BRACE SAW** in crotch for starting. *Don't maneuver with saw running.*

**3 TO MAKE CUT,** place your body weight in the saddle and brace yourself firmly with your feet.

**4 FINAL** limb-dropping cut is made while you're safely out of the way (preferably above) of the falling limb.

**5 TO AVOID DAMAGE,** tree man can aid ground man by using guide rope tied at balance point to help control limb.

**6 FELLING** larger branches can cause the tree to sway. Be prepared for this sudden movement by bracing yourself.

■ Check out your saw on the ground. Make sure it starts easily, and that the chain does not turn when the engine is idling.
■ Up in the tree, don't start the engine until you have positioned yourself to make the cut, and switch off the engine the moment the limb drops. Don't try to maneuver in the tree with the saw running.
■ Brace the saw firmly in a tree fork when you start it.
■ Keep a safety rope tied to the saw handle at its balance point. Have the other end of the safety rope secured or controlled by the ground man. Keep enough slack in the rope so that if you drop the saw, it will remain suspended several feet below any position you might slip to, yet not present any danger to someone on the ground. This way, if you lose control, you can drop the saw to use your hands to steady yourself without damaging the saw.
■ Be sure the safety rope on the saw and your own safety rope are away from the path of any movement of the limb rope when the limb drops or swings.

Always make sure your safety rope is properly attached and supported by a limb sound enough to hold your weight.

You can use an extension ladder to climb up into the branches. However, you may want it removed after you are in the tree to clear the way for lowering branches.

To get your safety rope across a suitable branch, it can be tossed over, using the method illustrated, or you can climb the ladder and put it over by hand. Where possible, the rope should be passed all the way around the trunk, using the crotch as a stop rather than a support. Be sure that the tree crotch is not so tight it prevents the rope from being pulled through easily.

Tie your saddle in one end of the rope and attach it to the tail rope, using the treeman's knot. The knot should be only a foot or so from your saddle knots, within easy reach of your hand.

Always test your saddle and treeman's knots before going up in the tree. You can do this by suspending yourself just above the ground.

**trimming hints**

There is no standard method for trimming a tree. Generally, trimmers start at the top and work down, dislodging caught branches as they descend. However, you may want to get some of the larger lower branches out of the way first.

When you saw off a branch, leave a stub one or two feet long to aid you in your climbing and rope handling. It can be cut off on your final descent if the entire tree is not to be taken down.

Remember, too, it's easy to underestimate the weight of a large limb, with all its twigs and leaves. It can be a very dangerous force. So, first cut off as many outer branches from the limb as possible and take the limb down in sections.

If you saw off a fairly large limb, the tree is relieved of considerable weight. This can cause the entire tree to sway back and forth quite violently. You must anticipate this and be prepared to switch off your chain saw immediately, and position it safely, while you brace yourself against the swaying.

Always leave a few feet of slack in the limb rope when you tie it off. There should be enough slack to let the limb drop below the working area, yet prevent it from coming down far enough to damage anything below.

# Tire chains—how to go in snow

**To keep moving in winter, snow tires are helpful, but tests by the National Safety Council show that chains are the way to go**

By BILL HARTFORD

■ SNOWBOUND in a toasty warm cabin, rubbing noses with your favorite movie star . . . it's a nice fantasy. But if you've got to be into your long johns and out on your appointed rounds—which is more likely the case—you must know a few things about going in snow.

Yes, it's cold and wet, and it can stop you dead in your tracks—especially if your tracks are not in the shape of snowshoes. Dog sleds, snowmobiles, maybe skis and snowshoes will take you to town, if you've got the stamina—and/or the dogs—but if you take the family car, and the family car is not a four-wheel-drive Jeep, then you'd better own a pair of tire chains and know how to use them.

Tire chains get their wicked grip from 12 or 13 cross chains that run across the treads (like radial plies) every several inches. The best of these cross chains have each link reinforced with a bar or lug, and it's this type that gives the best bite of all. The lugs, called V-bars on American Chain Co.'s Weed brand chains, are what really dig into the snow and ice. On dry pavement with your chains on, it's the lugs or bars that contact the surface, and they have to wear away completely before the links of cross chains even begin to wear. So they cost a little more than standard or twist-link types, but longer life and better bite make them worth it.

A pair of chains for the drive wheels of your car runs between $20 and $40, depending on your tire size—the larger the tire, the greater the number of links required and hence the higher the price. Slightly cheaper than steel-link chains are plastic traction treads. They're not as effective as steel, but they're lightweight, easier to handle and quieter on dry pavement. Both types eventually need repair and all manufacturers offer replacement cross-links and treads.

SEE ALSO
**Autos, gas mileage . . . Blizzards, driving . . . Noises, auto . . . Snowplows . . . Steering and suspension, auto . . . Tires . . . Winterizing, autos**

### Starting and pulling ability

**CHECKING** pulling power with strain gauge in the National Safety Council (NSC) tests.

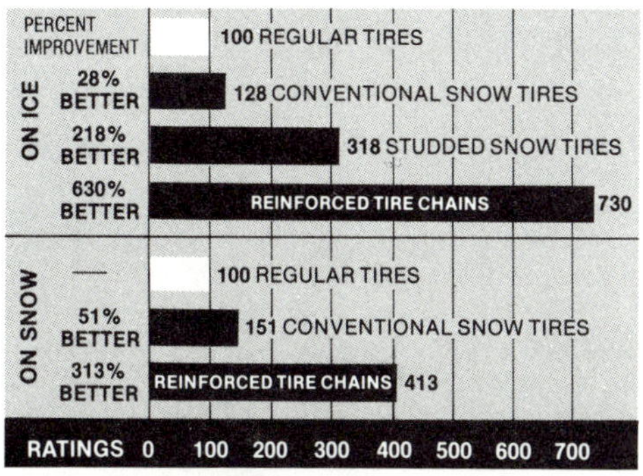


**PULLING ABILITY** of regular tires versus snow tires, studded snow tires and reinforced tire chains is compared above. The graphs, presented by the Safe Winter Driving League, were derived from tests conducted by the NSC.

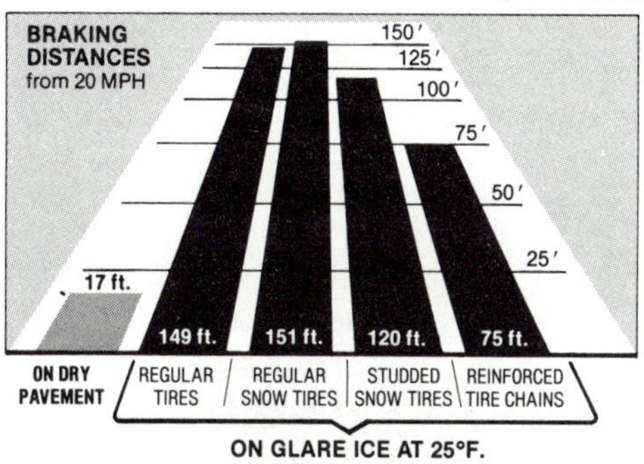


**ON GLARE ICE** at 25° F., stopping distance of a car with reinforced tire chains is half what it would be if the car were shoed with regular snow tires. Data dramatizes the fact that in winter the way to go—and the way to stop—is with tire chains.

#### TIRE CHAIN MANUFACTURERS

**Steel-Link Tire Chains**

American Chain & Cable Co., Inc., 935 Connecticut Ave., Bridgeport, CT 06602.
Campbell Chain Co., 3990 East Market St., York, PA 17402.
Nixdorff-Lloyd Chain Co., Dept. T, 1416 East 8th St., Winona, MN 55987.
Teledyne McKay, 850 Grantley Rd., York, PA 17405.
Western Chain Co., 1816 Belmont Ave., Chicago, IL 60657.

**Cable Chains**

Champion Cable & Chain Co., Box 162, Swanton, VT 05488.

**Elastomeric traction treads**

Detwiler Corp., 36 New York Ave., Westbury, NY 11590.

### Putting on chains without a jack

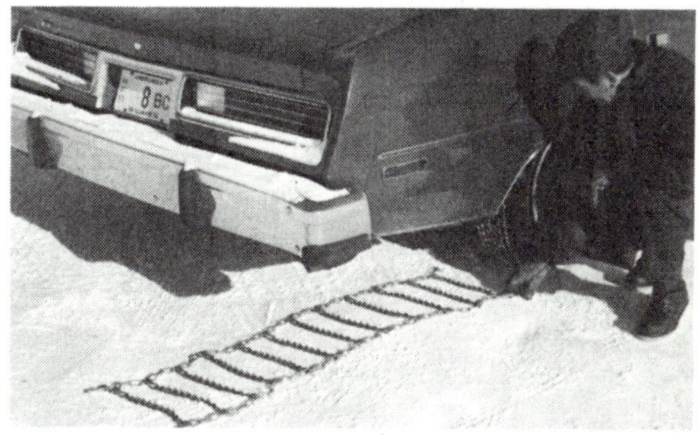

1 **LAY OUT CHAIN** straight behind car. Cross-chain hooks should face up, and with reinforced chains, lugs also should face up.

2 **GATHER CHAINS** behind the wheel, attach end links on spring steel chain applier, and slip the applier onto the tire.

3 **DRIVE THE CAR** forward until the applier is back to its starting point; the chain will now be wrapped around the tire.

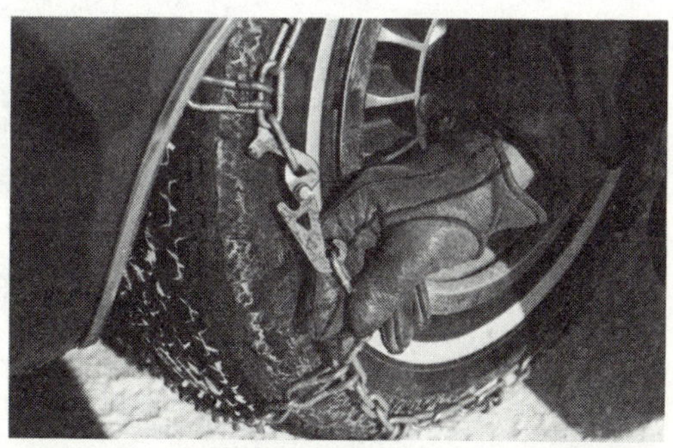

4 **REMOVE THE APPLIER;** fasten the inner lock first and then the outer lock into the proper links for a nice, snug fit.

**SPRAY-PAINTING SMALL KNOBS** or other small parts can be easily accomplished with this simple method. First, mount the knobs on a paper plate, using a screw or nail of the proper size. Support the plate underneath with one hand and slowly rotate it while spray-painting.—*Ken Patterson, Regina, Sask.*

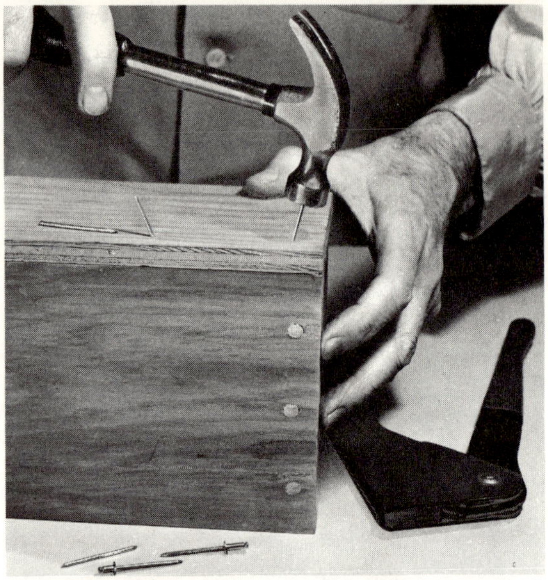

**AVOID WASTING** the sharp-pointed, snapped-off mandrels (stems) of Pop rivets. They can be reused as headless nails for many applications, including temporary assemblies and reinforcement of glued joints. For a stronger joint, it may help to drive successive stems at opposing angles.—*Walter E. Burton, Akron, OH.*

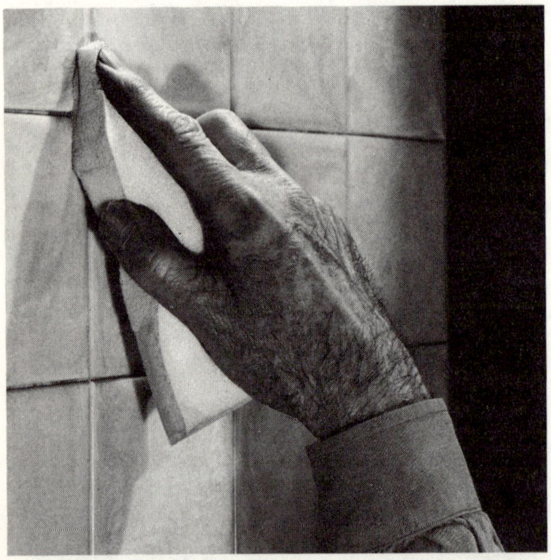

**A PIECE OF PLASTIC FOAM** can assist you a great deal in your grouting projects. For example, when grouting between plastic tiles, try a piece of 1-in.-thick plastic foam with one corner cut off. You can easily control the amount of grout applied and minimize the amount of excess grout deposited on the tiles.—*Paul J. Fister, Sac City, IA.*

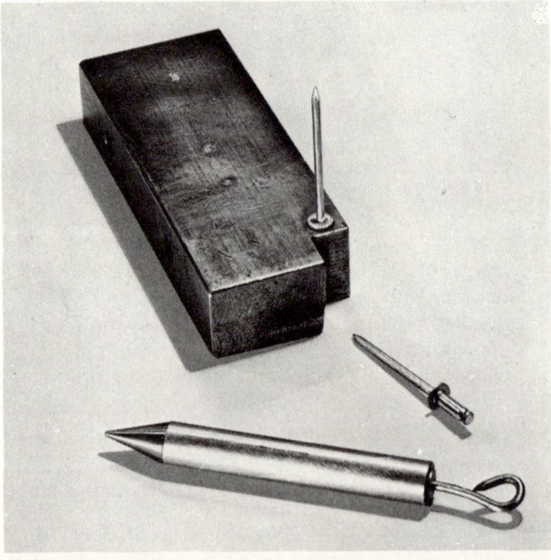

**AN EASY WAY** to install a nail-like stud in metal is to drill a hole to receive a Pop rivet, then expand the rivet only far enough to anchor it. The riveting tool is removed and the projecting stem is cut to length or bent as desired. An eye installed in a plumb bob using this technique will rotate freely. —*E. B. Walters, Chicago.*

# METRIC CONVERSION

Conversion factors can be carried so far they become impractical. In cases below where an entry is exact it is followed by an asterisk (*). Where considerable rounding off has taken place, the entry is followed by a + or a - sign.

## CUSTOMARY TO METRIC

### Linear Measure

| inches | millimeters |
|---|---|
| 1/16 | 1.5875* |
| 1/8 | 3.2 |
| 3/16 | 4.8 |
| 1/4 | 6.35* |
| 5/16 | 7.9 |
| 3/8 | 9.5 |
| 7/16 | 11.1 |
| 1/2 | 12.7* |
| 9/16 | 14.3 |
| 5/8 | 15.9 |
| 11/16 | 17.5 |
| 3/4 | 19.05* |
| 13/16 | 20.6 |
| 7/8 | 22.2 |
| 15/16 | 23.8 |
| 1 | 25.4* |

| inches | centimeters |
|---|---|
| 1 | 2.54* |
| 2 | 5.1 |
| 3 | 7.6 |
| 4 | 10.2 |
| 5 | 12.7* |
| 6 | 15.2 |
| 7 | 17.8 |
| 8 | 20.3 |
| 9 | 22.9 |
| 10 | 25.4* |
| 11 | 27.9 |
| 12 | 30.5 |

| feet | centimeters | meters |
|---|---|---|
| 1 | 30.48* | .3048* |
| 2 | 61 | .61 |
| 3 | 91 | .91 |
| 4 | 122 | 1.22 |
| 5 | 152 | 1.52 |
| 6 | 183 | 1.83 |
| 7 | 213 | 2.13 |
| 8 | 244 | 2.44 |
| 9 | 274 | 2.74 |
| 10 | 305 | 3.05 |
| 50 | 1524* | 15.24* |
| 100 | 3048* | 30.48* |

1 yard =
.9144* meters
1 rod =
5.0292* meters
1 mile =
1.6 kilometers
1 nautical mile =
1.852* kilometers

### Fluid Measure

(Milliliters [ml] and cubic centimeters [cc or cu cm] are equivalent, but it is customary to use milliliters for liquids.)

1 cu in = 16.39 ml
1 fl oz = 29.6 ml
1 cup = 237 ml
1 pint = 473 ml
1 quart = 946 ml
= .946 liters
1 gallon = 3785 ml
= 3.785 liters
Formula (exact):
fluid ounces × 29.573 529 562 5*
= milliliters

### Weights

| ounces | grams |
|---|---|
| 1 | 28.3 |
| 2 | 56.7 |
| 3 | 85 |
| 4 | 113 |
| 5 | 142 |
| 6 | 170 |
| 7 | 198 |
| 8 | 227 |
| 9 | 255 |
| 10 | 283 |
| 11 | 312 |
| 12 | 340 |
| 13 | 369 |
| 14 | 397 |
| 15 | 425 |
| 16 | 454 |

Formula (exact):
ounces × 28.349 523 125* =
grams

| pounds | kilograms |
|---|---|
| 1 | .45 |
| 2 | .9 |
| 3 | 1.4 |
| 4 | 1.8 |
| 5 | 2.3 |
| 6 | 2.7 |
| 7 | 3.2 |
| 8 | 3.6 |
| 9 | 4.1 |
| 10 | 4.5 |

1 short ton (2000 lbs) =
907 kilograms (kg)
Formula (exact):
pounds × .453 592 37* =
kilograms

### Volume

1 cu in = 16.39 cubic centimeters (cc)
1 cu ft = 28 316.7 cc
1 bushel = 35 239.1 cc
1 peck = 8 809.8 cc

### Area

1 sq in = 6.45 sq cm
1 sq ft = 929 sq cm
= .093 sq meters
1 sq yd = .84 sq meters
1 acre = 4 046.9 sq meters
= .404 7 hectares
1 sq mile = 2 589 988 sq meters
= 259 hectares
= 2.589 9 sq kilometers

### Kitchen Measure

1 teaspoon = 4.93 milliliters (ml)
1 Tablespoon = 14.79 milliliters (ml)

### Miscellaneous

1 British thermal unit (Btu) (mean)
= 1 055.9 joules
1 calorie (mean) = 4.19 joules
1 horsepower = 745.7 watts
= .75 kilowatts
caliber (diameter of a firearm's bore in hundredths of an inch)
= .254 millimeters (mm)
1 atmosphere pressure = 101 325* pascals (newtons per sq meter)
1 pound per square inch (psi) = 6 895 pascals
1 pound per square foot = 47.9 pascals
1 knot = 1.85 kilometers per hour
25 miles per hour = 40.2 kilometers per hour
50 miles per hour = 80.5 kilometers per hour
75 miles per hour = 120.7 kilometers per hour